Unstoppable Owl

Name: Date: Score: Unstoppable Owl

1) Identifying Fractions

Fractions are used to identify parts of a whole. How many parts does the circle in the example have? It has 4 equal parts. How many of those parts are shaded? 3 of the 4 parts are shaded. So, you could say that **three-fourths** of the sections are shaded. Three-fourths can also be written as the fraction $\frac{3}{4}$. 3 is the **numerator** and 4 is the **denominator**.

Example:

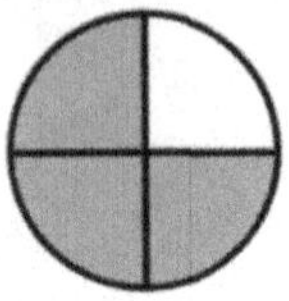 $= \frac{3}{4}$

3 ← The number of parts that are shaded (the **numerator**).

4 ← The number of equal parts there are (the **denominator**).

Directions: *How many parts are shaded? Write the numerator* ***above*** *the fraction's line.*

1) 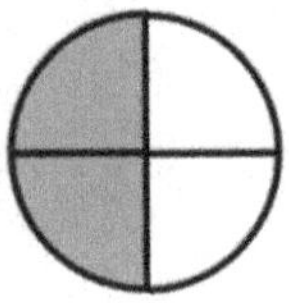 $= \frac{}{4}$

2) 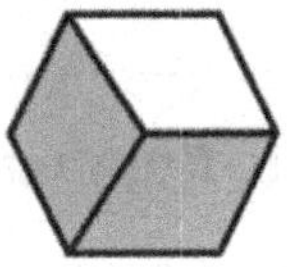 $= \frac{}{3}$

3) 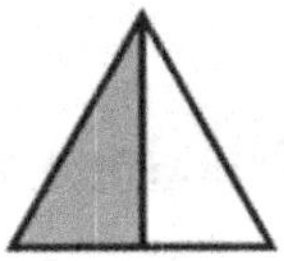 $= \frac{}{2}$

4) 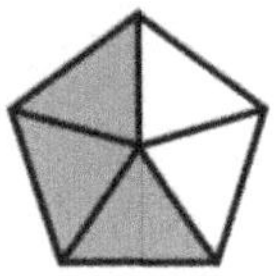 $= \frac{}{5}$

5) 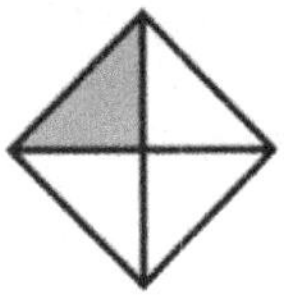 $= \frac{}{4}$

6)  $= \frac{}{4}$

7)  $= \frac{}{6}$

8)  $= \frac{}{3}$

9) 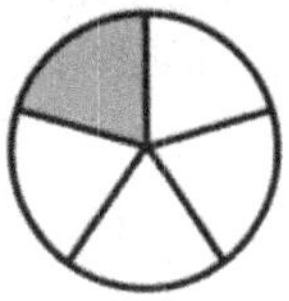 $= \frac{}{5}$

Directions: *How many equal parts are there? Write the denominator* ***below*** *the fraction's line.*

10)  $= \frac{2}{}$

11) 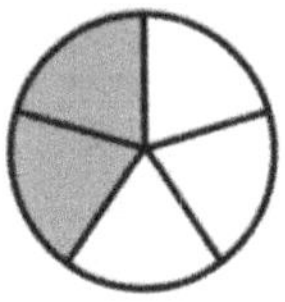 $= \frac{2}{}$

12)  $= \frac{3}{}$

13) 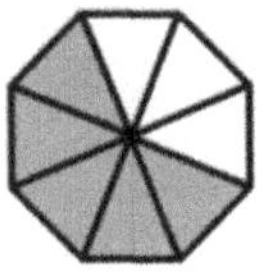 $= \frac{5}{}$

14) 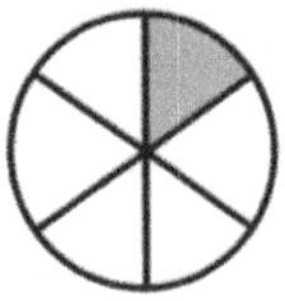 $= \frac{1}{}$

15)  $= \frac{1}{}$

16) 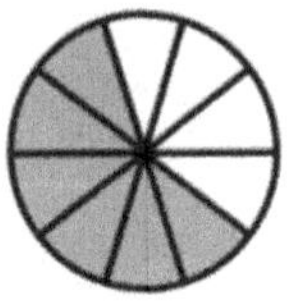 $= \frac{6}{}$

17) 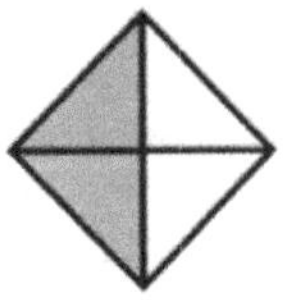 $= \frac{2}{}$

18)  $= \frac{4}{}$

Name: Date: Score: Unstoppable Owl

2) Identifying Fractions

Directions: *How many parts are shaded? Write the numerator **above** the fraction's line.*

1) 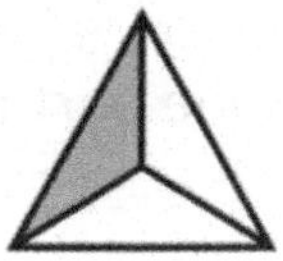 $= \frac{\quad}{3}$

2)  $= \frac{\quad}{5}$

3) 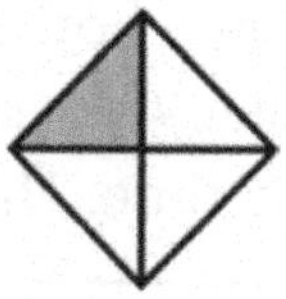 $= \frac{\quad}{4}$

4) 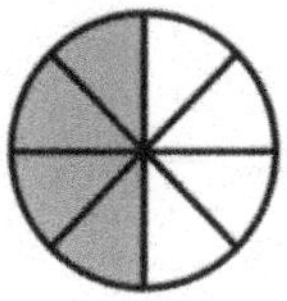 $= \frac{\quad}{8}$

5)  $= \frac{\quad}{2}$

6)  $= \frac{\quad}{4}$

Directions: *How many equal parts are there? Write the denominator **below** the fraction's line.*

7) 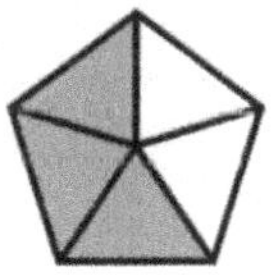 $= \frac{3}{\quad}$

8) 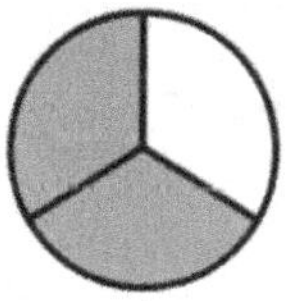 $= \frac{2}{\quad}$

9)  $= \frac{1}{\quad}$

10)  $= \frac{2}{\quad}$

11) 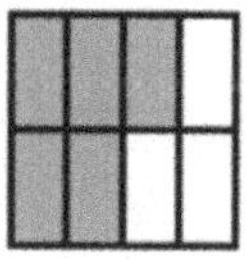 $= \frac{5}{\quad}$

12) 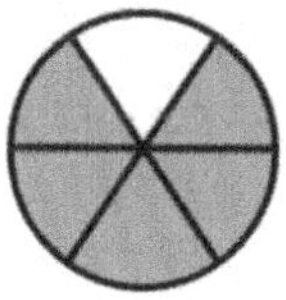 $= \frac{5}{\quad}$

Mixed Review: *Write the missing part of the fraction.*

13) 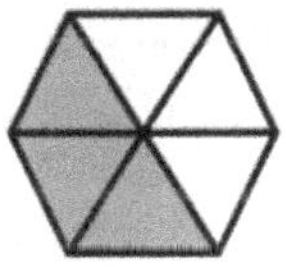 $= \frac{3}{\quad}$

14) 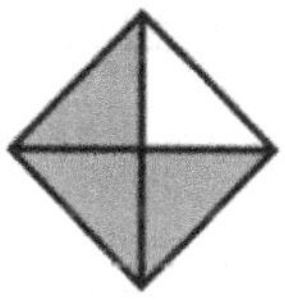 $= \frac{\quad}{4}$

15) 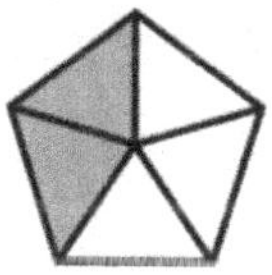 $= \frac{2}{\quad}$

16) 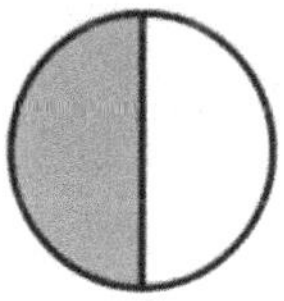 $= \frac{\quad}{2}$

17) 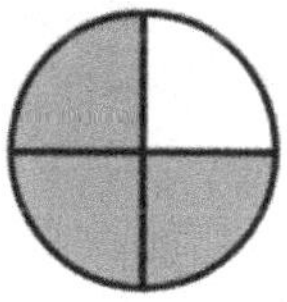 $= \frac{\quad}{4}$

18) 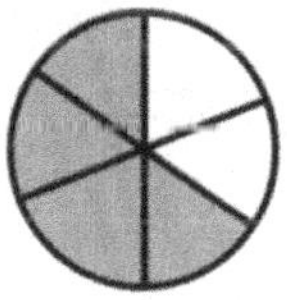 $= \frac{4}{\quad}$

19)  $= \frac{1}{\quad}$

20)  $= \frac{4}{\quad}$

21)  $= \frac{\quad}{2}$

22)  $= \frac{4}{\quad}$

23) 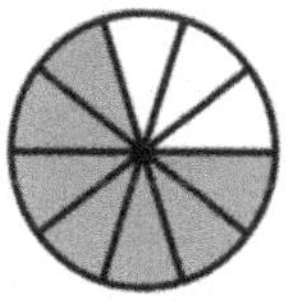 $= \frac{7}{\quad}$

24) 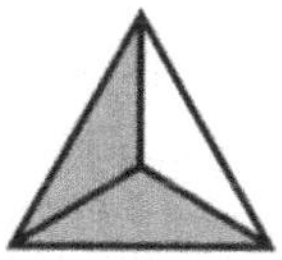 $= \frac{\quad}{3}$

Name: Date: Score: Unstoppable Owl

3) Identifying Fractions

You've practiced finding numerators, and you've practiced finding denominators. Now, it's time to identify both the numerator and denominator.

Work carefully. Ask yourself what each part of the fraction means. Focus on one part of the fraction at a time. Then, write the fraction.

Directions: *Write the fraction that is shown.*

1) 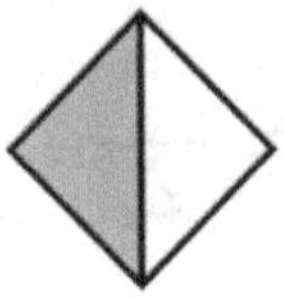 = —

2) 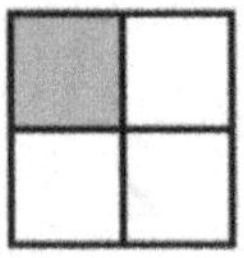 = —

3) 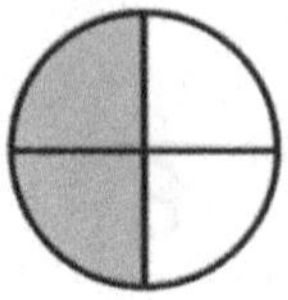 = —

4) 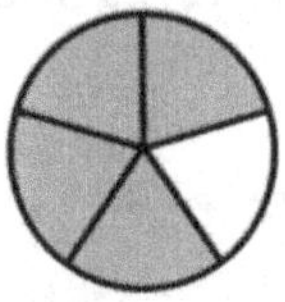 = —

5)  = —

6) 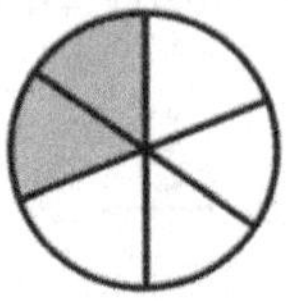 = —

7)  = —

8) 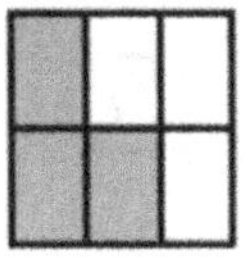 = —

9) 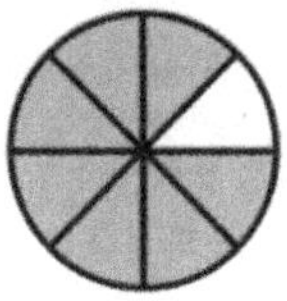 = —

10) 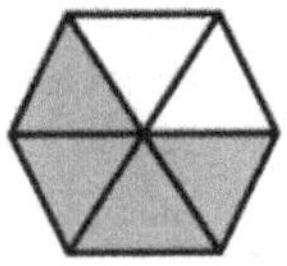 = —

11) 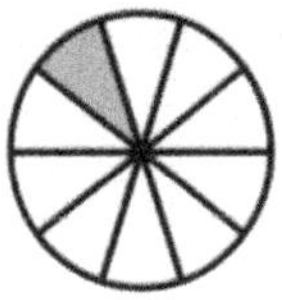 = —

12) 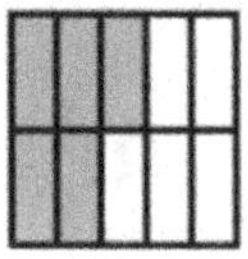 = —

13) 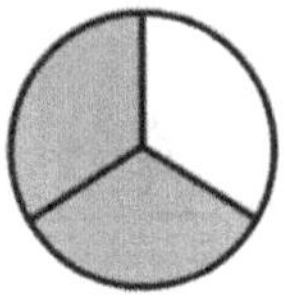 = —

14)  = —

15) 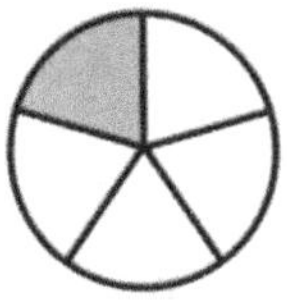 = —

16) 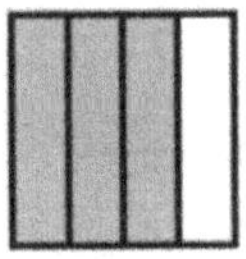 = —

17) 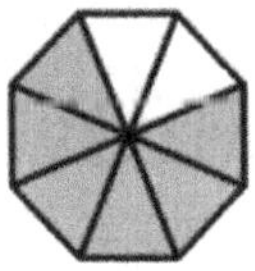 = —

18) 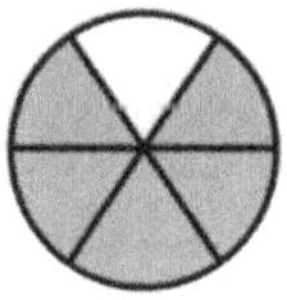 = —

19)  = —

20) 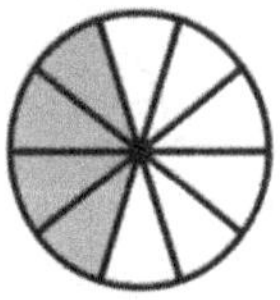 = —

21) 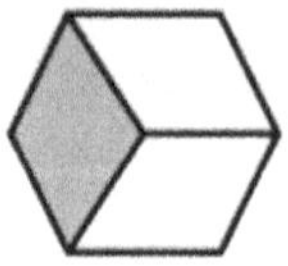 = —

Name: Date: Score: Unstoppable Owl

4) Identifying Fractions

Fractions can be written in **number form**, but they can also be written in **word form**.

Study the examples to the right. Say the name of each fraction out loud.

Number Form:	Word Form:
$\frac{3}{4}$	Three-fourths
$\frac{1}{2}$	One-half
$\frac{2}{3}$	Two-thirds
$\frac{4}{5}$	Four-fifths

Directions: *Draw a line to match the number form of each fraction with its word form.*

$\frac{1}{5}$ 1) A. **Four-ninths**

$\frac{3}{6}$ 2) B. **Six-sevenths**

$\frac{4}{9}$ 3) C. **One-fifth**

$\frac{1}{4}$ 4) D. **Three-sixths**

$\frac{6}{7}$ 5) E. **One-fourth**

$\frac{5}{12}$ 6) A. **One-half**

$\frac{1}{2}$ 7) B. **Seven-eighths**

$\frac{7}{8}$ 8) C. **Four-tenths**

$\frac{2}{6}$ 9) D. **Five-twelfths**

$\frac{4}{10}$ 10) E. **Two-sixths**

Directions: *Write each fraction in* ***word form.***

Number Form:	Word Form:
$\frac{2}{4}$	11)
$\frac{1}{3}$	12)
$\frac{5}{6}$	13)
$\frac{4}{7}$	14)
$\frac{1}{5}$	15)
$\frac{3}{10}$	16)
$\frac{5}{8}$	17)
$\frac{3}{5}$	18)
$\frac{2}{9}$	19)
$\frac{7}{12}$	20)

Write each fraction in ***number form.***

Number Form:	Word Form:
21)	One-sixth
22)	Three-eighths
23)	Two-fifths
24)	Five-ninths
25)	One-half
26)	Two-sevenths
27)	Six-elevenths
28)	One-tenth
29)	Seven-eighths
30)	Four-fifths

Name: Date: Score:

Unstoppable Owl

5) Identifying Fractions

Directions: *Write the fraction that is shown.*

1)  = —

2) 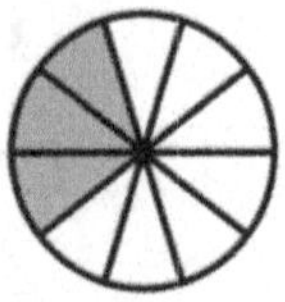 = —

3) 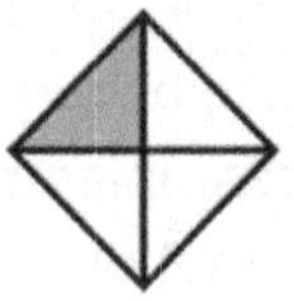 = —

4) 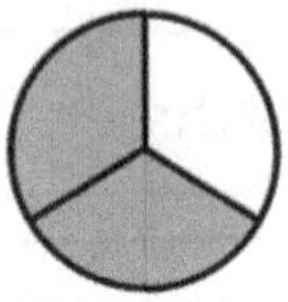 = —

5)  = —

6)  = —

7) 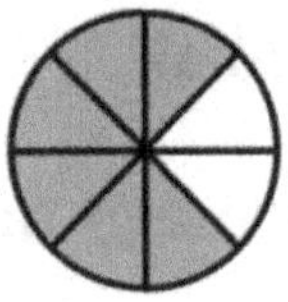 = —

8)  = —

9)  = —

10)  = —

11) 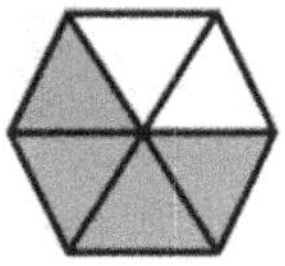 = —

12) 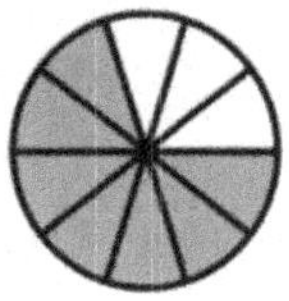 = —

Directions: *Write each fraction in **word form**.*

Number Form:	Word Form:
$\frac{6}{7}$	13)
$\frac{4}{9}$	14)
$\frac{5}{11}$	15)
$\frac{2}{3}$	16)
$\frac{1}{4}$	17)
$\frac{2}{6}$	18)
$\frac{9}{10}$	19)
$\frac{3}{8}$	20)
$\frac{1}{2}$	21)
$\frac{2}{5}$	22)

*Write each fraction in **number form**.*

Number Form:	Word Form:
23)	**Three-ninths**
24)	**One-third**
25)	**Five-sevenths**
26)	**Eight-tenths**
27)	**Three-fourths**
28)	**One-fifth**
29)	**Four-eighths**
30)	**Three-sixths**
31)	**Five-twelfths**
32)	**One-seventh**

Name: Date: Score: Unstoppable Owl

6) Identifying Fractions

Directions: *Shade the illustration to represent each fraction. (There may be multiple correct ways to shade each illustration. Just make sure the right number of parts are shaded.)*

1) 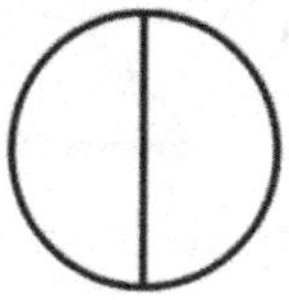 $= \frac{1}{2}$

2) 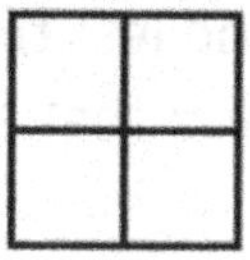 $= \frac{2}{4}$

3) 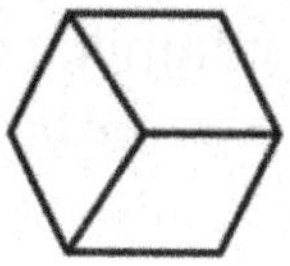 $= \frac{1}{3}$

4)  $= \frac{7}{8}$

5) 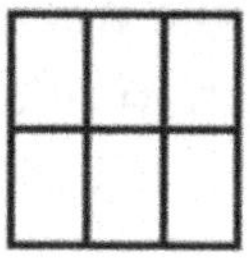 $= \frac{3}{6}$

6)  $= \frac{2}{5}$

7)  $= \frac{3}{10}$

8)  $= \frac{1}{4}$

9) 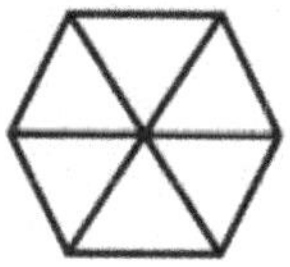 $= \frac{4}{6}$

10)  $= \frac{2}{3}$

11) 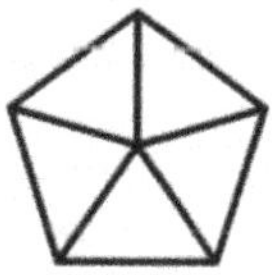 $= \frac{4}{5}$

12)  $= \frac{9}{10}$

13) 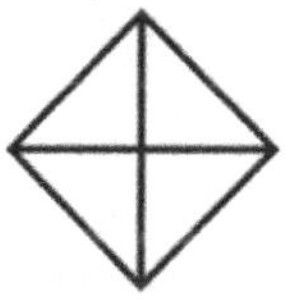 $= \frac{3}{4}$

14)  $= \frac{2}{6}$

15)  $= \frac{1}{5}$

16) 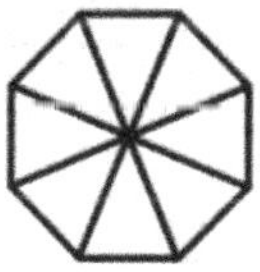 $= \frac{5}{8}$

17) 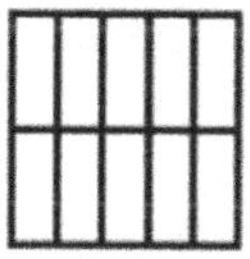 $= \frac{7}{10}$

18) 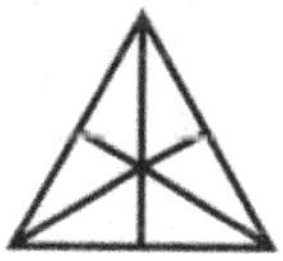 $= \frac{5}{6}$

19)  $= \frac{3}{5}$

20) 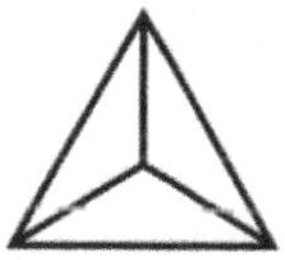 $= \frac{1}{3}$

21)  $= \frac{3}{4}$

22) 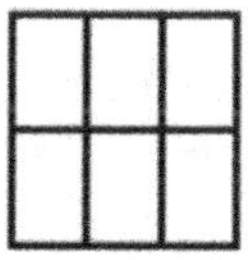 $= \frac{1}{6}$

23)  $= \frac{4}{5}$

24)  $= \frac{5}{10}$

Name: Date: Score:

7) Comparing Fractions (With Like Denominators)

You may be familiar with comparing whole numbers. For instance, **9 is greater than 2**. This can be written as **9 > 2**.

If you are not familiar with the greater than and less than symbols, they are easy to learn. People often say that the opening of the symbol is like a hungry mouth, that wants to eat the larger number. The point of the symbol points at the number that has a lesser value.

9 > 2

Nine **is greater than** two.

(*The teeth are imaginary.*)

Fractions can be compared the same way whole numbers are compared. Some fractions have a greater value than others. The greater than and less than symbols can be used to compare fractions too. An example using pictures is shown, as well as an example when the fractions are written in number form.

Examples:

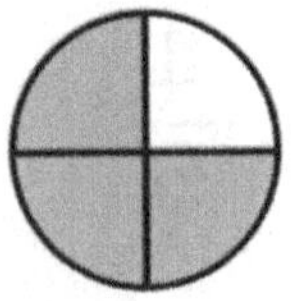 > 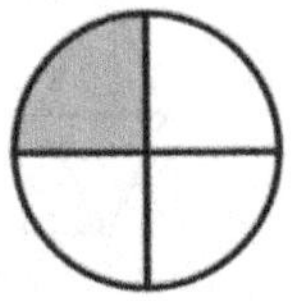

Three-fourths **is greater than** one-fourth.

$\frac{3}{5} < \frac{4}{5}$

Three-fifths **is less than** four-fifths.

Directions: *Draw a greater than or less than symbol to compare each set of fractions.*

1) 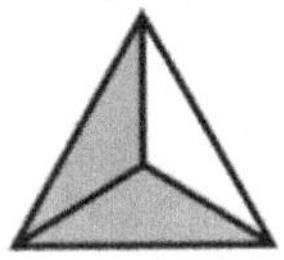 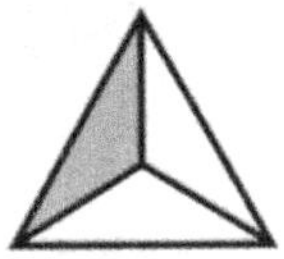

2) 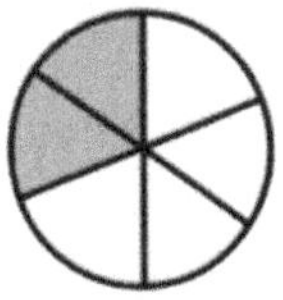 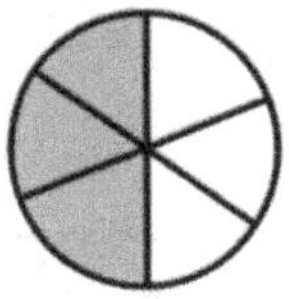

3) 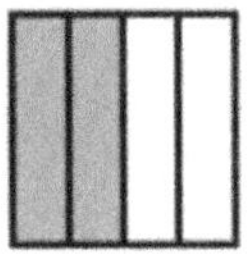 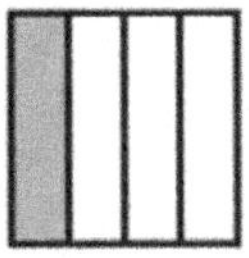

4) 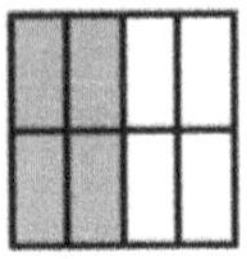 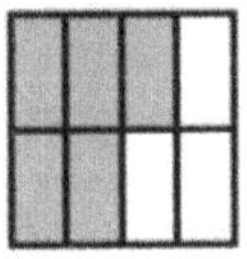

5) 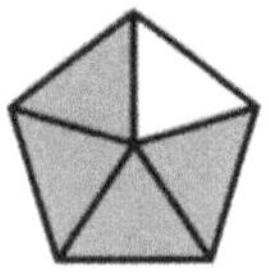 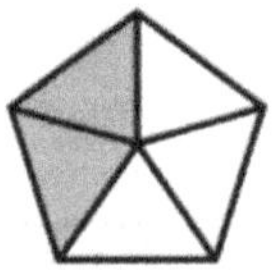

6) 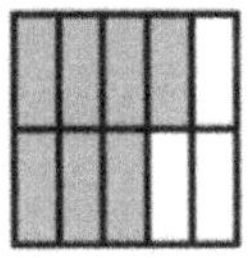 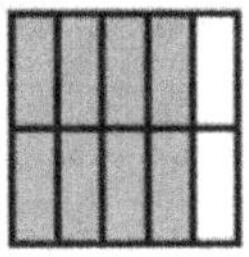

7) $\frac{2}{4}$ $\frac{3}{4}$

8) $\frac{2}{7}$ $\frac{1}{7}$

9) $\frac{1}{5}$ $\frac{3}{5}$

10) $\frac{5}{6}$ $\frac{4}{6}$

11) $\frac{7}{10}$ $\frac{5}{10}$

12) $\frac{3}{8}$ $\frac{6}{8}$

13) $\frac{1}{3}$ $\frac{2}{3}$

14) $\frac{8}{9}$ $\frac{7}{9}$

15) $\frac{10}{12}$ $\frac{11}{12}$

Name: Date: Score: Unstoppable Owl

8) Comparing Fractions (With Unlike Denominators)

When fractions have like denominators, the fraction that has the greatest numerator will have the greatest value. But that is not always the case for fractions that have unlike denominators. Comparing fractions with unlike denominators requires a little more thinking.

The examples below have an illustration to represent each fraction. In the first example, three-eighths is being compared with two-fourths. Three-eighths may have the larger numerator, but since each eighth is smaller than each of the fourths, two-fourths still has a greater value. You can even see in the illustration that more of the circle is shaded for two-fourths than is shaded for three-eighths. Two-fourths is greater.

Examples:

$\frac{3}{8} < \frac{2}{4}$

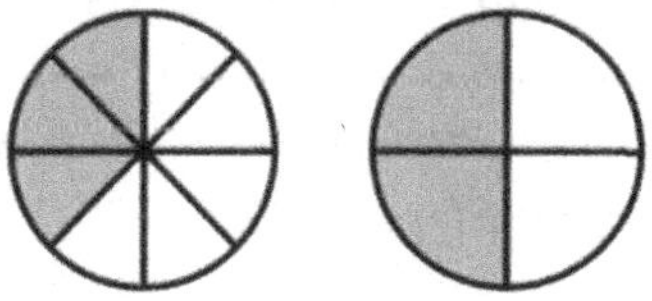

Three-eighths **is less than** two-fourths.

$\frac{2}{3} > \frac{2}{10}$

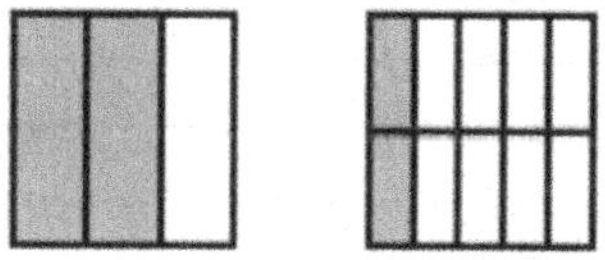

Two-thirds **is greater than** two-tenths.

Directions: *Draw a greater than or less than symbol to compare each set of fractions.*

1) $\frac{3}{6}$ $\frac{2}{3}$

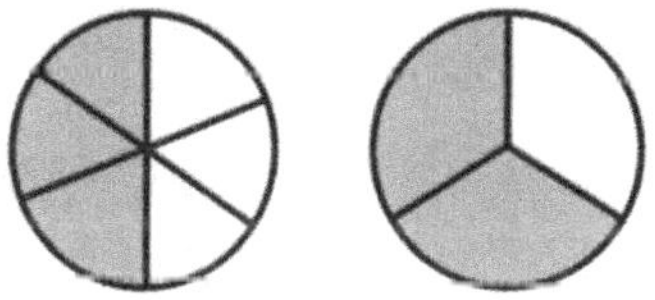

2) $\frac{4}{5}$ $\frac{1}{3}$

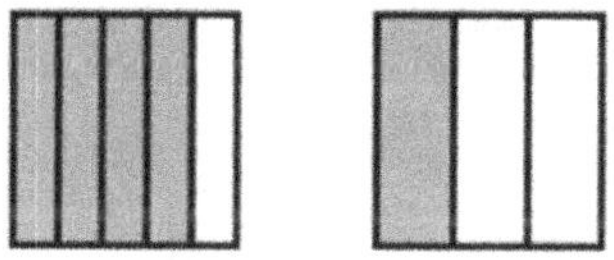

3) $\frac{1}{2}$ $\frac{3}{10}$

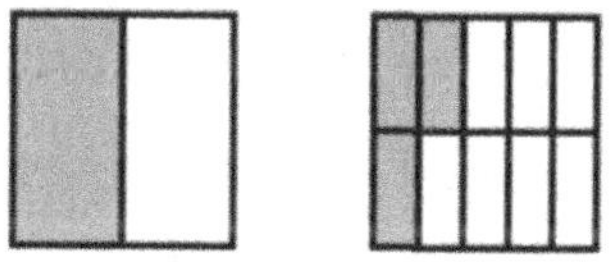

4) $\frac{1}{6}$ $\frac{3}{8}$

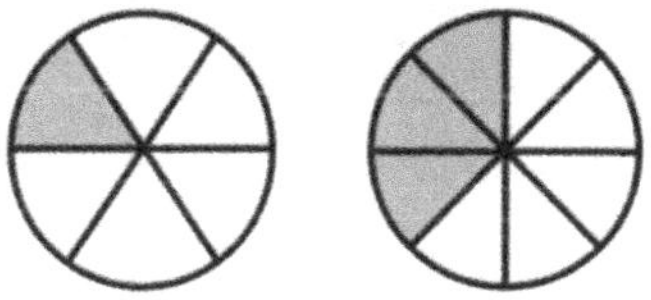

5) $\frac{4}{8}$ $\frac{1}{4}$

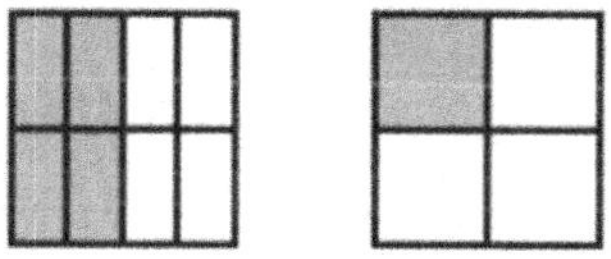

6) $\frac{2}{5}$ $\frac{2}{3}$

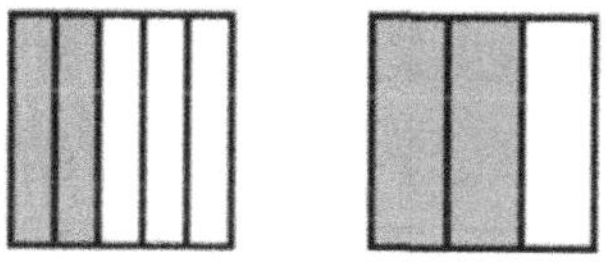

7) $\frac{5}{10}$ $\frac{4}{6}$

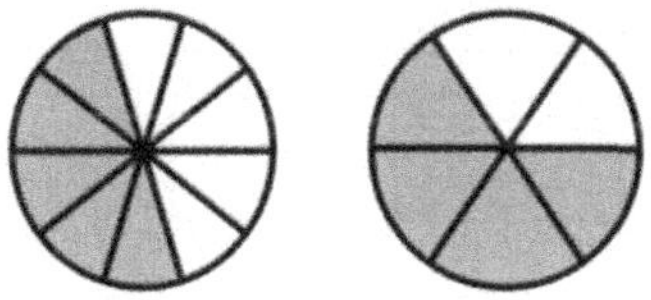

8) $\frac{2}{3}$ $\frac{1}{2}$

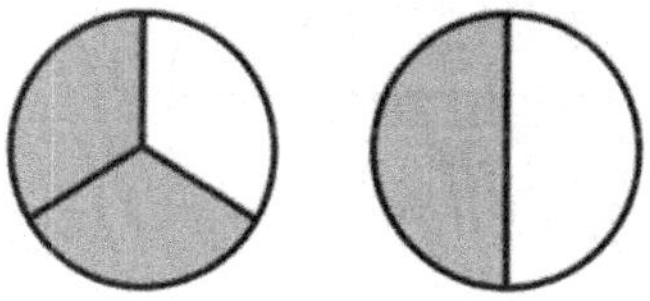

9) $\frac{4}{8}$ $\frac{3}{4}$

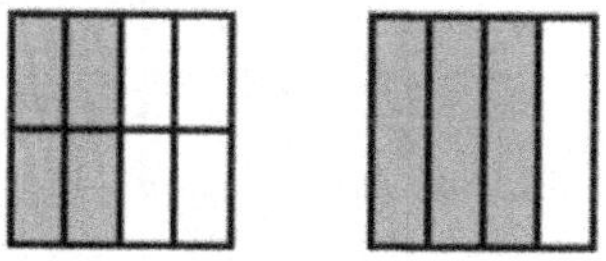

Name: Date: Score: Unstoppable Owl

9) Comparing Fractions (With Unlike Denominators)

A **fraction bar** is a type of illustration that can be useful when comparing fractions. When two fraction bars are placed one on top of the other, it's easy to see which fraction has the greater value. The bar that is shaded further has the greatest value.

Examples:

$\frac{2}{4} < \frac{4}{7}$

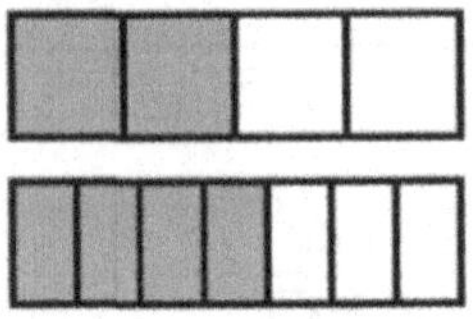

Two-fourths **is less than** four-sevenths.

$\frac{1}{3} > \frac{3}{10}$

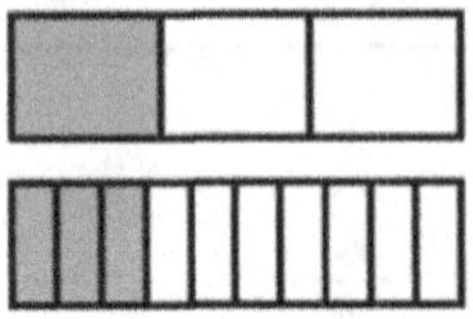

One-third **is greater than** three-tenths.

Directions: *Draw a greater than or less than symbol to compare each set of fractions.*

1) $\frac{1}{2}$ $\frac{3}{5}$

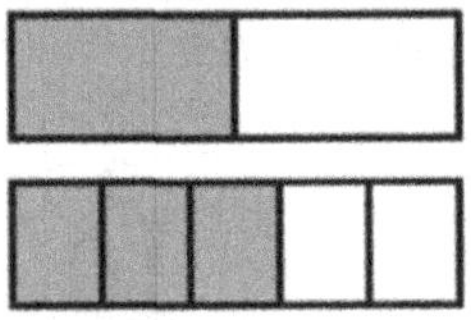

2) $\frac{3}{4}$ $\frac{4}{7}$

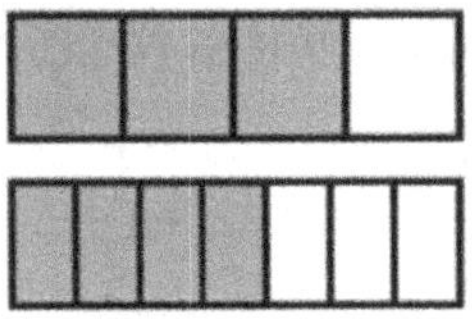

3) $\frac{2}{4}$ $\frac{5}{8}$

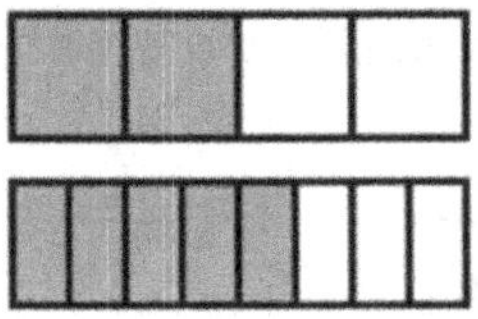

4) $\frac{1}{6}$ $\frac{1}{10}$

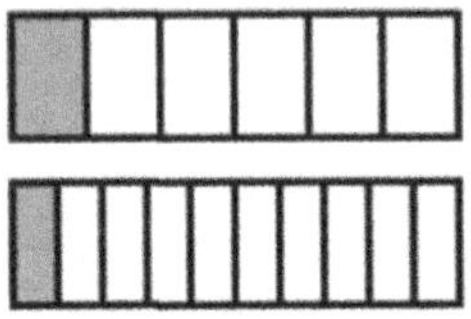

5) $\frac{3}{5}$ $\frac{4}{6}$

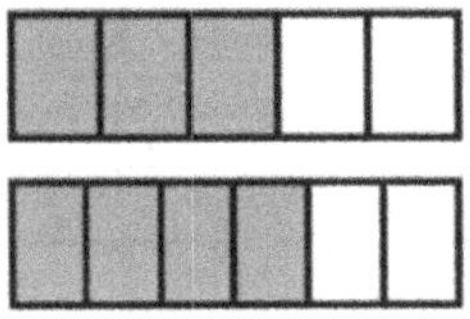

6) $\frac{8}{10}$ $\frac{7}{8}$

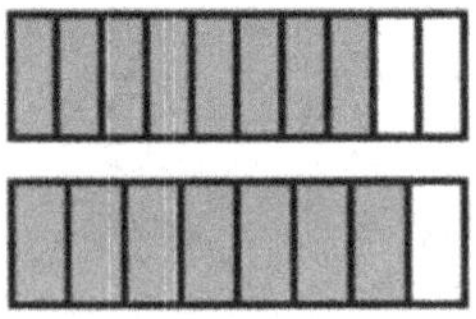

7) $\frac{3}{4}$ $\frac{4}{5}$

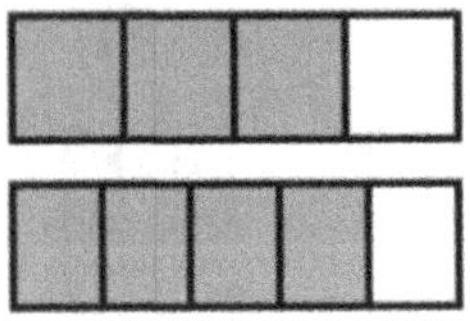

8) $\frac{7}{10}$ $\frac{2}{3}$

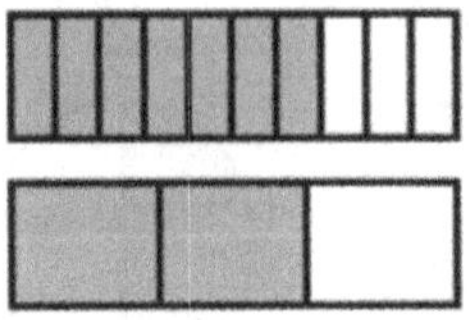

9) $\frac{5}{9}$ $\frac{1}{2}$

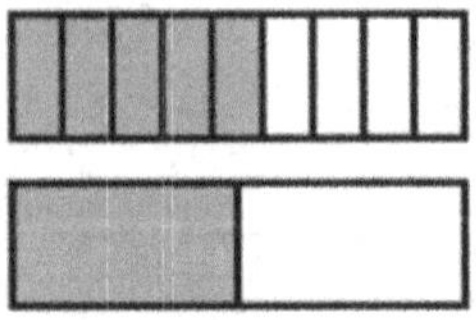

Name: Date: Score: Unstoppable Owl

10) Comparing Fractions (Equivalent Fractions)

When comparing fractions, one fraction does not need to be greater than or less than the other fraction. When different fractions have equal values, they're called **equivalent fractions**. An **equal sign** (=) can be used to compare equivalent fractions.

One-half and two-fourths are equivalent fractions. They are different fractions, but their values are equal. One is not greater than or less than the other.

Examples:

$\frac{1}{2} = \frac{2}{4}$

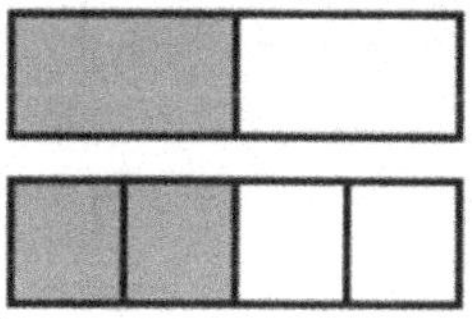

One-half **is equal to** two-fourths.

$\frac{4}{6} = \frac{2}{3}$

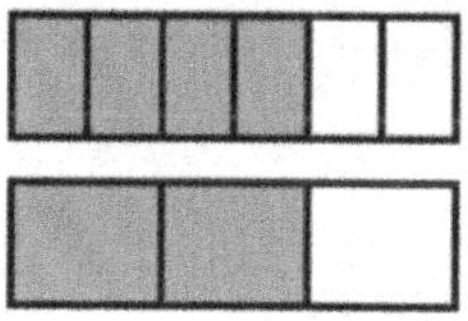

Four-sixths **is equal to** two-thirds.

Directions: *Draw a <, >, or = symbol to compare each set of fractions.*

1) $\frac{1}{4}$ $\frac{2}{8}$

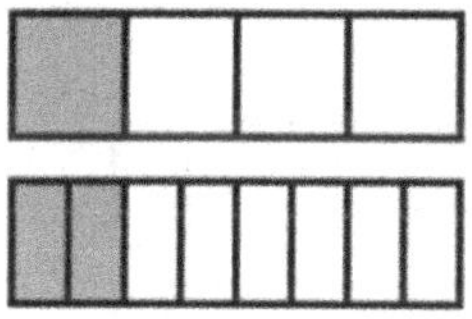

2) $\frac{3}{4}$ $\frac{6}{7}$

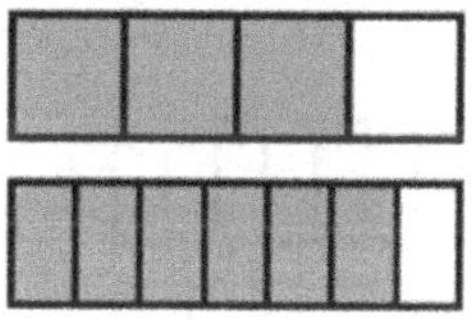

3) $\frac{6}{9}$ $\frac{2}{3}$

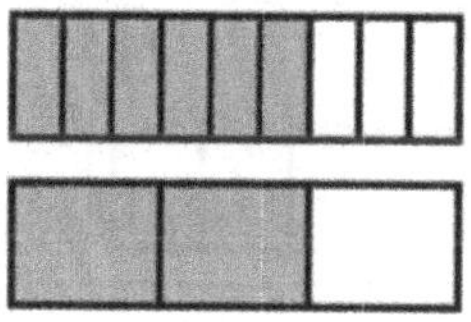

4) $\frac{3}{5}$ $\frac{1}{2}$

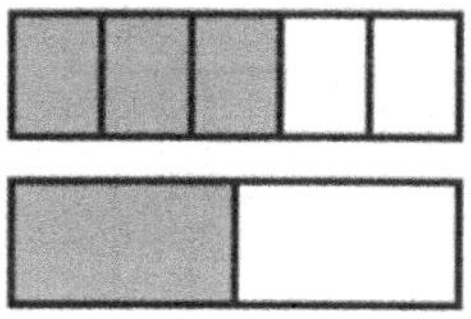

5) $\frac{2}{4}$ $\frac{5}{10}$

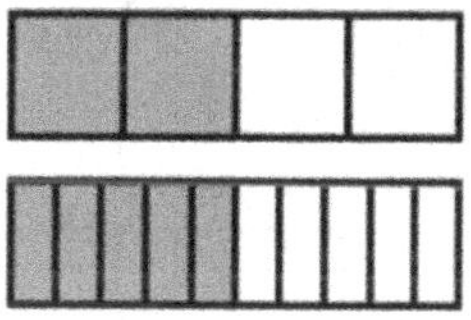

6) $\frac{2}{5}$ $\frac{4}{9}$

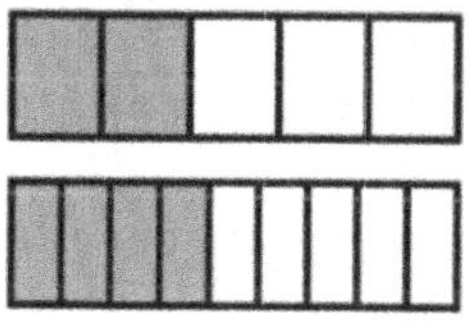

7) $\frac{7}{10}$ $\frac{4}{6}$

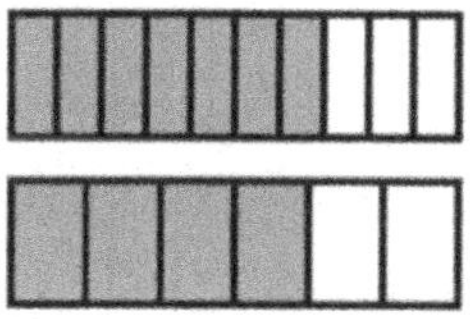

8) $\frac{3}{4}$ $\frac{6}{8}$

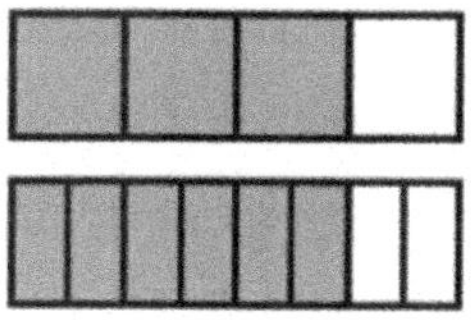

9) $\frac{2}{10}$ $\frac{1}{5}$

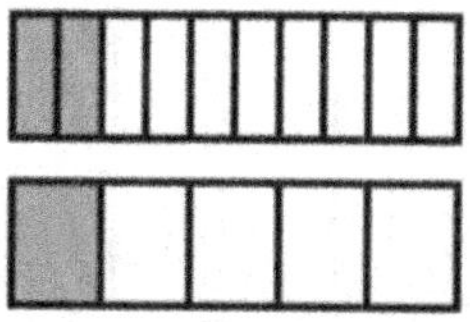

Unstoppable Owl

Name: Date: Score:

11) Comparing Fractions (With Unlike Denominators)

On this page, blank fraction bars are provided, but you will need to shade them. Start on the left side of the fraction bars and shade the spaces needed to represent each fraction.

Directions: *Draw a <, >, or = symbol to compare each set of fractions.*

1) $\frac{2}{3}$ $\frac{3}{4}$

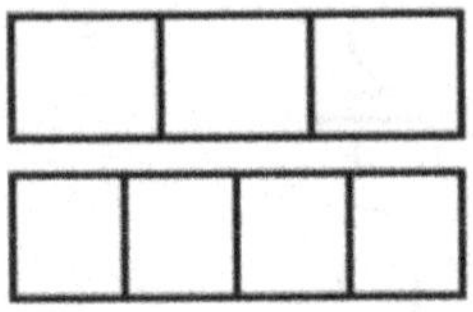

2) $\frac{3}{5}$ $\frac{4}{7}$

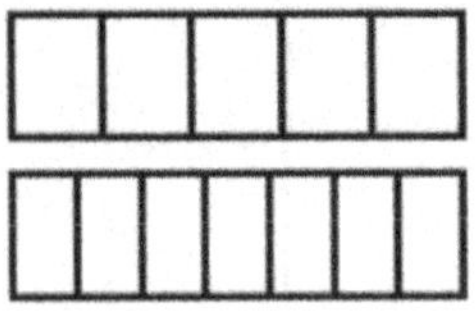

3) $\frac{1}{2}$ $\frac{2}{5}$

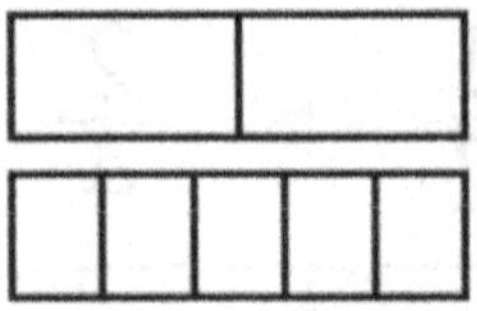

4) $\frac{4}{8}$ $\frac{5}{10}$

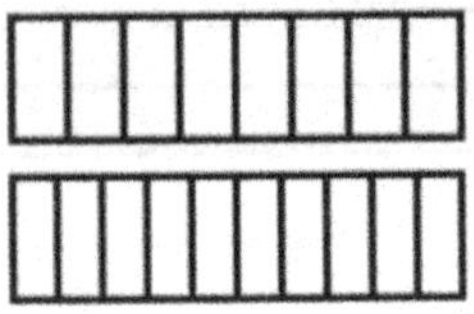

5) $\frac{2}{6}$ $\frac{1}{4}$

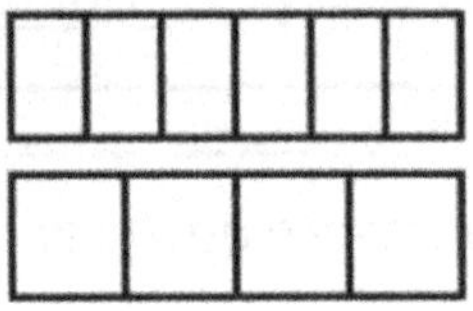

6) $\frac{1}{3}$ $\frac{2}{7}$

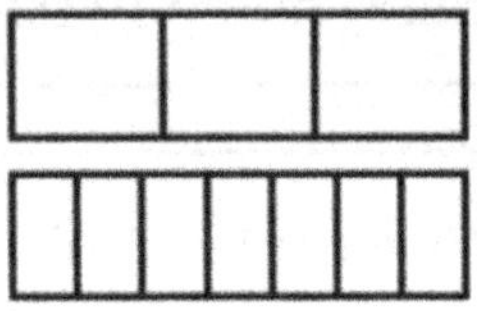

7) $\frac{4}{5}$ $\frac{6}{8}$

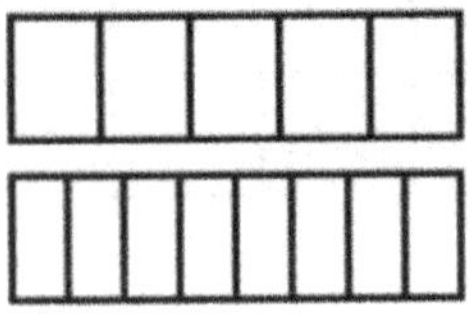

8) $\frac{3}{9}$ $\frac{2}{6}$

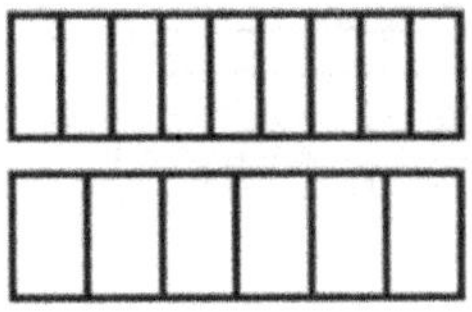

9) $\frac{4}{6}$ $\frac{2}{3}$

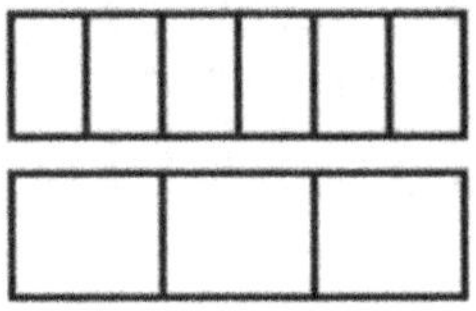

10) $\frac{1}{2}$ $\frac{5}{9}$

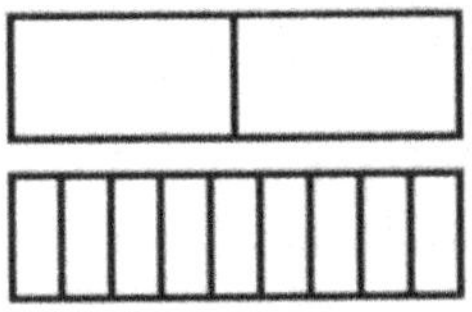

11) $\frac{1}{5}$ $\frac{2}{8}$

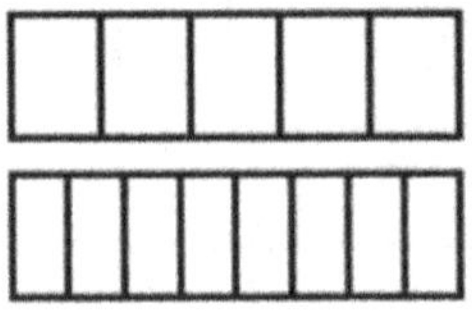

12) $\frac{5}{10}$ $\frac{3}{6}$

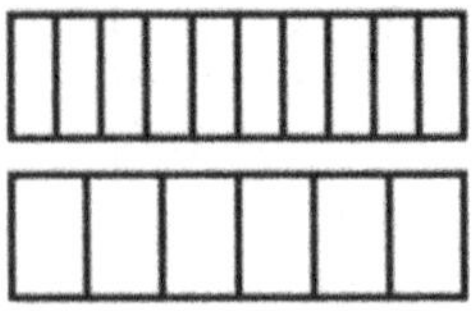

13) $\frac{3}{4}$ $\frac{6}{8}$

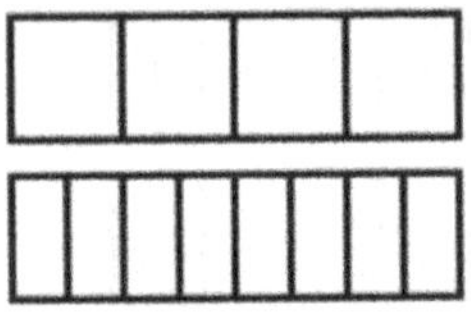

14) $\frac{3}{8}$ $\frac{1}{3}$

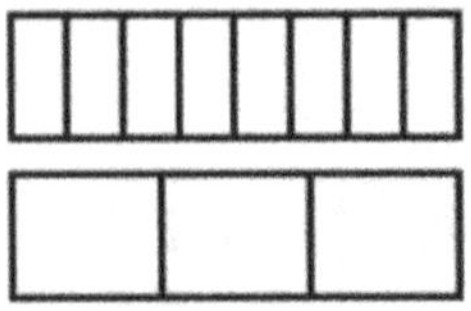

15) $\frac{7}{9}$ $\frac{6}{7}$

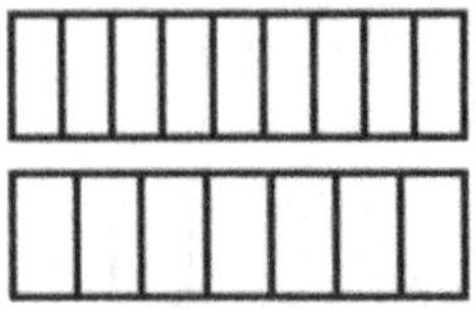

Name: Date: Score: *Unstopppable Owl*

12) Comparing Fractions (Without Illustrations)

Illustrations are not needed to compare fractions. With practice, you can roughly visualize representations of a fraction in your mind. Visualization is often enough to estimate whether one fraction is greater or less than another.

Directions: *Draw a <, >, or = symbol to compare each set of fractions.*

1) $\frac{1}{4}$ $\frac{1}{2}$

2) $\frac{2}{6}$ $\frac{4}{5}$

3) $\frac{3}{4}$ $\frac{2}{3}$

4) $\frac{2}{4}$ $\frac{5}{10}$

5) $\frac{2}{5}$ $\frac{1}{2}$

6) $\frac{2}{8}$ $\frac{1}{4}$

7) $\frac{1}{3}$ $\frac{4}{7}$

8) $\frac{3}{8}$ $\frac{8}{9}$

9) $\frac{2}{10}$ $\frac{3}{6}$

10) $\frac{1}{2}$ $\frac{2}{3}$

11) $\frac{3}{4}$ $\frac{3}{7}$

12) $\frac{2}{6}$ $\frac{4}{8}$

13) $\frac{3}{5}$ $\frac{2}{4}$

14) $\frac{1}{3}$ $\frac{1}{7}$

15) $\frac{3}{4}$ $\frac{7}{8}$

16) $\frac{3}{6}$ $\frac{1}{3}$

17) $\frac{1}{4}$ $\frac{3}{8}$

18) $\frac{9}{10}$ $\frac{2}{3}$

Challenge: *Write each set of fractions in order from least to greatest.*

19) $\frac{1}{2}$ $\frac{2}{3}$ $\frac{5}{6}$ $\frac{3}{9}$

20) $\frac{1}{3}$ $\frac{2}{4}$ $\frac{3}{5}$ $\frac{2}{10}$

Name: Date: Score: Unstoppable Owl

13) Comparing Fractions (Making Equivalent Fractions)

The key to making equivalent fractions is to make sure that they have the same value. Multiplication can be used to do this. If you multiply the numerator and denominator with the same number, an equivalent fraction will be found. This is called a multiplication factor.

Examples:

$$\frac{1 \times 2}{3 \times 2} = \frac{\mathbf{2}}{\mathbf{6}} \qquad \frac{3 \times 5}{5 \times 5} = \frac{\mathbf{15}}{\mathbf{25}} \qquad \frac{2 \times 4}{3 \times 4} = \frac{\mathbf{8}}{\mathbf{12}}$$

Directions: *Multiply to make an equivalent fraction.*

1) $\frac{2 \times 3}{5 \times 3} = \frac{\_\_}{\_\_}$ 2) $\frac{1 \times 4}{4 \times 4} = \frac{\_\_}{\_\_}$ 3) $\frac{6 \times 2}{7 \times 2} = \frac{\_\_}{\_\_}$

4) $\frac{1 \times 5}{2 \times 5} = \frac{\_\_}{\_\_}$ 5) $\frac{2 \times 3}{3 \times 3} = \frac{\_\_}{\_\_}$ 6) $\frac{1 \times 6}{5 \times 6} = \frac{\_\_}{\_\_}$

When only one number of an equivalent fraction is known, division or a combination of division and multiplication may often be used to find the unknown numerator or the unknown denominator. If the unknown fraction has a larger number than the known fraction, both division and multiplication may be used. If not, only division may be needed.

Examples:	**Step 1:** Use division	**Step 2:** × or ÷	**Answer:**
$\frac{3}{4} = \frac{\_\_}{20}$	20 ÷ 4 = **5**	3 × 5 = **15**	$\frac{3}{4} = \frac{\mathbf{15}}{20}$
$\frac{8}{28} = \frac{2}{\_\_}$	8 ÷ 2 = **4**	28 ÷ 4 = **7**	$\frac{8}{28} = \frac{2}{\mathbf{7}}$

Directions: *Write the missing numbers to create equivalent fractions.*

7) $\frac{4}{5} = \frac{\_\_}{10}$ 8) $\frac{8}{12} = \frac{2}{\_\_}$ 9) $\frac{6}{18} = \frac{\_\_}{9}$

10) $\frac{15}{40} = \frac{3}{\_\_}$ 11) $\frac{6}{7} = \frac{30}{\_\_}$ 12) $\frac{8}{16} = \frac{\_\_}{2}$

Name: Date: Score: Unstoppable Owl

14) Comparing Fractions (Making Equivalent Fractions)

Directions: *Write the missing numbers to create equivalent fractions.*

1) $\frac{30}{40} = \frac{6}{}$

2) $\frac{24}{30} = \frac{}{10}$

3) $\frac{3}{4} = \frac{}{16}$

4) $\frac{4}{7} = \frac{20}{}$

5) $\frac{6}{8} = \frac{3}{}$

6) $\frac{12}{36} = \frac{}{6}$

7) $\frac{5}{9} = \frac{}{27}$

8) $\frac{20}{100} = \frac{2}{}$

9) $\frac{4}{6} = \frac{}{24}$

10) $\frac{35}{50} = \frac{}{10}$

11) $\frac{3}{9} = \frac{12}{}$

12) $\frac{10}{25} = \frac{2}{}$

Challenge: *Circle the one fraction in each set that is not an equivalent fraction to the others.*

13) $\frac{4}{8}$ $\frac{8}{12}$ $\frac{1}{2}$ $\frac{5}{10}$

14) $\frac{6}{9}$ $\frac{4}{6}$ $\frac{10}{15}$ $\frac{3}{4}$

15) $\frac{3}{10}$ $\frac{6}{15}$ $\frac{2}{5}$ $\frac{8}{20}$

16) $\frac{3}{12}$ $\frac{2}{8}$ $\frac{4}{14}$ $\frac{4}{16}$

17) $\frac{15}{20}$ $\frac{8}{12}$ $\frac{3}{4}$ $\frac{18}{24}$

18) $\frac{8}{18}$ $\frac{3}{6}$ $\frac{15}{30}$ $\frac{6}{12}$

19) $\frac{3}{9}$ $\frac{5}{15}$ $\frac{2}{6}$ $\frac{4}{10}$

20) $\frac{8}{10}$ $\frac{6}{8}$ $\frac{12}{15}$ $\frac{20}{25}$

Name: Date: Score:

15) Fractions & Decimals (Tenths)

Fraction		*Decimal*
$\frac{2}{10}$	=	0.2

This is the **decimal point.**

Fractions are not the only way parts of a whole number can be written. **Decimals** are often used too. Fractions and decimals can be used to represent the same value. They just represent it differently. Fractions are written with two numbers that are separated by a line. Decimals are written using a **decimal point**.

The **tenths** place is the first number to the right of a decimal point. It represents a part of a whole. For example, 0.7 is seven-tenths and 0.4 is four-tenths. Each of these decimals can be written as a fraction. Notice that the denominator of the fractions are tenths as well. These fractions and decimals are the same number, representing the same value.

Examples:

 $= \frac{7}{10} = 0.7$

 $= \frac{4}{10} = 0.4$

Directions: *Write the decimal for each fraction.*

1)  $= \frac{3}{10} =$

2)  $= \frac{8}{10} =$

3)  $= \frac{6}{10} =$

4)  $= \frac{2}{10} =$

5)  $= \frac{1}{10} =$

6)  $= \frac{4}{10} =$

7)  $= \frac{9}{10} =$

8)  $= \frac{5}{10} =$

Name: Date: Score: Unstoppable Owl

16) Fractions & Decimals (Tenths)

Directions: *Write the decimal for each fraction.*

1)  =

2)  =

3)  =

4)  =

5)  =

6)  =

7)  =

8) 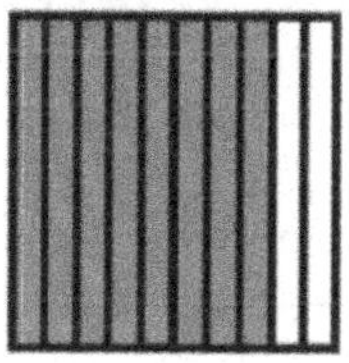 =

9)  =

10) $\frac{4}{10}$ =

11) $\frac{8}{10}$ =

12) $\frac{1}{10}$ =

13) $\frac{6}{10}$ =

14) $\frac{9}{10}$ =

15) $\frac{3}{10}$ =

16) $\frac{7}{10}$ =

17) $\frac{5}{10}$ =

Directions: *Write the fraction for each decimal.*

18) — = 0.2

19) — = 0.9

20) — = 0.6

21) — = 0.1

22) — = 0.7

23) — = 0.5

24) — = 0.3

25) — = 0.8

Name: Date: Score:

17) Fractions & Decimals (Hundredths)

Fraction	Decimal
$\frac{27}{100}$ =	0.27
Tenths	Hundredths

There can be more than one number behind a decimal point. A second number behind the decimal point represents hundredths. If one whole is divided into one hundred equal parts, like the squares in these illustrations are, each of the one hundred parts would be worth one hundredth of the whole. If 27 of the parts are shaded, then 27 hundredths of the whole are shaded.

The first number to the right of a decimal point is the tenths place. The second number is called the hundredths place. When written as a fraction, hundredths are represented with a 100 as the denominator.

Examples:

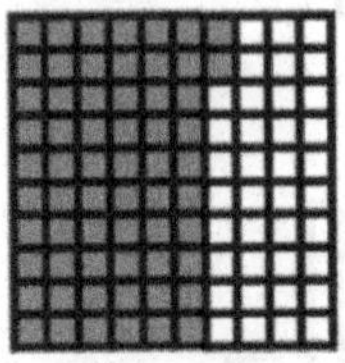 $= \frac{62}{100} = 0.62$

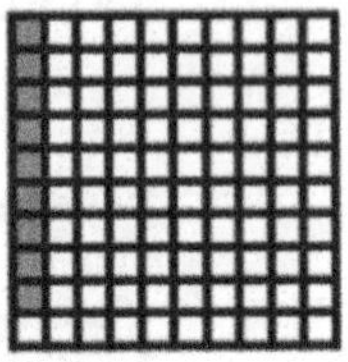 $= \frac{9}{100} = 0.09$

Directions: *Write the decimal for each fraction.*

1) 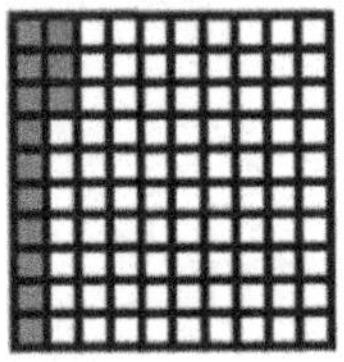 $= \frac{13}{100} =$

2)  $= \frac{41}{100} =$

3) 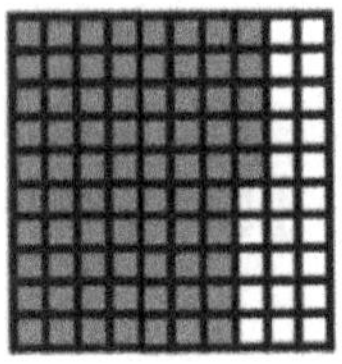 $= \frac{75}{100} =$

4)  $= \frac{94}{100} =$

5) 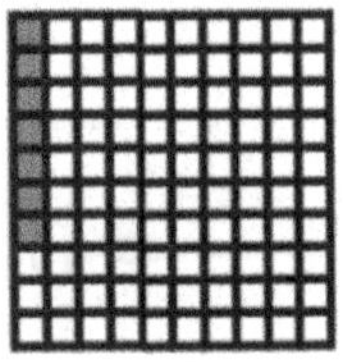 $= \frac{7}{100} =$

6)  $= \frac{59}{100} =$

7)  $= \frac{88}{100} =$

8) 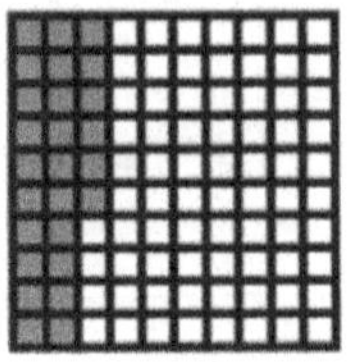 $= \frac{26}{100} =$

Name: Date: Score: *Unstoppable Owl*

18) Fractions & Decimals (Hundredths)

Directions: *Write the decimal for each fraction.*

1)  =

2)  =

3) 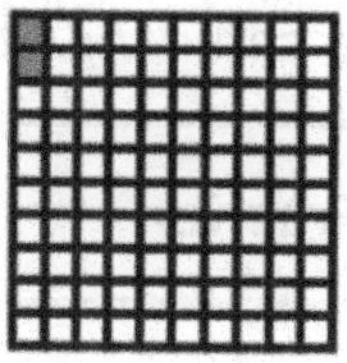 =

4)  =

5)  =

6)  =

7) 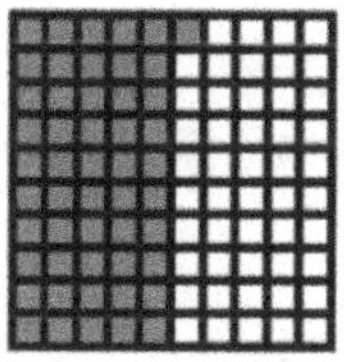 =

8)  =

9) 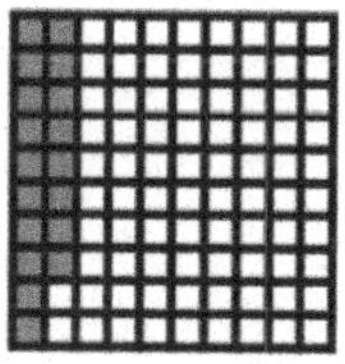 =

10) $\frac{57}{100} =$

11) $\frac{11}{100} =$

12) $\frac{92}{100} =$

13) $\frac{35}{100} =$

14) $\frac{84}{100} =$

15) $\frac{49}{100} =$

16) $\frac{78}{100} =$

17) $\frac{3}{100} =$

Directions: *Write the fraction for each decimal.*

18) ―― = 0.41

19) ―― = 0.28

20) ―― = 0.06

21) ―― = 0.94

22) ―― = 0.72

23) ―― = 0.85

24) ―― = 0.53

25) ―― = 0.17

Name: Date: Score:

Unstoppable Owl

19) Fractions & Decimals (Decimal Place Values)

Every fraction has a decimal equivalent. To find the decimal for each fraction, you can divide the numerator by the denominator.

Fraction	Divide the numerator by the denominator	Decimal
$\frac{3}{4}$	$3 \div 4 =$	0.75
$\frac{2}{5}$	$2 \div 5 =$	0.4
$\frac{13}{25}$	$13 \div 25 =$	0.52

Directions: *Divide to convert each fraction to a decimal. (A calculator may be used.)*

1) $\frac{1}{2} =$
2) $\frac{1}{4} =$
3) $\frac{4}{5} =$
4) $\frac{9}{15} =$
5) $\frac{13}{20} =$
6) $\frac{34}{40} =$
7) $\frac{12}{30} =$
8) $\frac{21}{25} =$
9) $\frac{39}{50} =$
10) $\frac{2}{5} =$
11) $\frac{14}{20} =$
12) $\frac{6}{40} =$

Tenths and hundredths are not the only fractions that can be represented by a decimal. Sometimes, there are more than two digits to the right of a decimal. Each extra digit represents its own place value.

Study the names of the place values in the example shown. *Do you recognize any similarities to the names of the place values for whole numbers*?

529.83647

- 5 — Hundreds
- 2 — Tens
- 9 — Ones
- 8 — Tenths
- 3 — Hundredths
- 6 — Thousandths
- 4 — Ten Thousandths
- 7 — Hundred Thousandths

Directions: *Name the place value of the underlined digit.*

13) $0.68\underline{2}94$
14) $0.1\underline{3}9$
15) $0.\underline{4}6$
16) $0.719\underline{5}8$
17) $0.84\underline{0}276$
18) $0.3856\underline{2}71$
19) $0.205\underline{7}49$
20) $0.947\underline{1}$

Name: Date: Score: Unstoppable Owl

20) Fractions & Decimals (Rounding Decimals)

Sometimes, there are more than two digits to the right of a decimal. In fact, the number of digits after a decimal may never end. These are called **non-terminating** decimals.

How do you write all those digits? Most of the time, people don't. They round the decimal.

How do you round a decimal? Rounding a decimal is similar to rounding a whole number. Determine what place value you are rounding to. Then, look at the digit to the right of that place value. If the digit is less than 5, round down. If the digit is 5 or greater, round up.

Decimal	Rounded to the nearest...			
	Tenth	Hundredth	Thousandth	Ten Thousandth
0.73572124	0.7	0.74	0.736	0.7357
0.1284799	0.1	0.13	0.128	0.1285
0.371802	0.4	0.37	0.372	0.3718
0.888888	0.9	0.89	0.889	0.8889

Directions: *Fill in the blank spaces by rounding each decimal.*

Decimal	Rounded to the nearest...			
	Tenth	Hundredth	Thousandth	Ten Thousandth
1) 0.372185				
2) 0.9357184				
3) 0.128753				
4) 0.555555				
5) 0.37136				
6) 0.7728901				
7) 0.481539				
8) 0.941768				
9) 0.222222				
10) 0.561893				
11) 0.6183517				
12) 0.86274				
13) 0.314725				
14) 0.76859				
15) 0.8417526				

Name: Date: Score: Unstoppable Owl

21) Fractions & Decimals (Repeating Decimals)

Directions: *Divide to convert each fraction to a decimal. (A calculator may be used.) When applicable, round long decimals to the nearest **hundredths** place.*

1) $\frac{3}{8}$ =
2) $\frac{9}{80}$ =
3) $\frac{7}{32}$ =
4) $\frac{5}{12}$ =

5) $\frac{29}{400}$ =
6) $\frac{17}{40}$ =
7) $\frac{9}{17}$ =
8) $\frac{11}{15}$ =

9) $\frac{323}{500}$ =
10) $\frac{27}{75}$ =
11) $\frac{7}{16}$ =
12) $\frac{125}{200}$ =

Non-terminating decimals may form a pattern of repeating numbers. This pattern keeps repeating indefinitely. When this happens, the decimal is called a **repeating decimal**.

Instead of writing a pattern repeatedly, people often abbreviate the repeating decimals by writing the repeating numbers once, and then draw a line over those numbers to indicate that they repeat.

Repeating Decimal	Abbreviation
0.4545454545...	$0.\overline{45}$
0.77777777...	$0.\overline{7}$
0.30818181...	$0.30\overline{81}$
0.156156156...	$0.\overline{156}$
0.802020202...	$0.8\overline{02}$

Directions: *Write each repeating decimal in its abbreviated form.*

13) 0.88888...
14) 0.474747...
15) 0.255555...

16) 0.5393939...
17) 0.126126...
18) 0.797979...

19) 0.643333...
20) 0.818181...
21) 0.22222...

22) 0.401401...
23) 0.9474747...
24) 0.582582...

Directions: *Divide to convert each fraction to a decimal. (A calculator may be used.) When applicable, write each repeating decimal in its abbreviated form.*

25) $\frac{2}{3}$ =
26) $\frac{17}{18}$ =
27) $\frac{6}{11}$ =
28) $\frac{11}{12}$ =

29) $\frac{5}{27}$ =
30) $\frac{5}{36}$ =
31) $\frac{13}{33}$ =
32) $\frac{8}{24}$ =

33) $\frac{10}{11}$ =
34) $\frac{15}{44}$ =
35) $\frac{3}{22}$ =
36) $\frac{21}{55}$ =

Name: Date: Score: Unstoppable Owl

22) Fractions & Decimals (Removing Trailing Zeros)

When a decimal ends with a zero, it is called a **trailing zero**. Many people do not write trailing zeros when writing a decimal. For example, instead of writing 0.70, they write 0.7.

Why would someone not write a trailing zero? A main reason is because trailing zeros can lead to errors. Many healthcare workers are prohibited from writing trailing zeros because they can be accidentally read as a 6 instead of a 0. Studies have shown that removing trailing zeros leads to fewer errors.

Trailing Zeros	No Trailing Zeros
0.70	0.7
0.1600	0.16
0.020	0.02
0.8000	0.8
0.7190	0.719
0.058000	0.058

Directions: *Rewrite each decimal so that it does not have a trailing zero.*

1) 0.30
2) 0.400
3) 0.760
4) 0.0810
5) 0.0050
6) 0.20
7) 0.04090
8) 0.3100
9) 0.27880
10) 0.70000
11) 0.02500
12) 0.00050
13) 0.800200
14) 0.2610
15) 0.5900
16) 0.000140
17) 0.94000
18) 0.00300
19) 0.387000
20) 0.1500
21) 0.0209040
22) 0.77030
23) 0.00080
24) 0.01000

Directions: *Round each decimal to the nearest thousandth and write it without a trailing zero.*

25) 0.7395
26) 0.509721
27) 0.29984
28) 0.0997
29) 0.499711
30) 0.93995
31) 0.009801
32) 0.1298
33) 0.6097
34) 0.5797
35) 0.89981
36) 0.0495

Name: Date: Score: Unstoppable Owl

23) Fractions & Decimals (Common Conversions)

Division can be used to convert a fraction to a decimal, but many times, division is not necessary. Memorizing common conversions between fractions and decimals can save you a lot of time and effort. With a little practice, conversions can be made mentally without needing to reach for a calculator.

Study the chart to the right. Then, try to convert between fractions and decimals without looking at the chart. Skip the problems you do not know. Once you are done, refer to the chart again to complete the problems. (You may want to cover the chart with a sheet of paper while you are working.)

Common Conversions

Fraction	Decimal
$\frac{1}{2}$	0.5
$\frac{1}{3}$	$0.\overline{3}$
$\frac{2}{3}$	$0.\overline{6}$
$\frac{1}{4}$	0.25
$\frac{2}{4}$	0.5
$\frac{3}{4}$	0.75
$\frac{1}{5}$	0.2
$\frac{2}{5}$	0.4
$\frac{3}{5}$	0.6
$\frac{4}{5}$	0.8
$\frac{1}{10}$	0.1
$\frac{2}{10}$	0.2
$\frac{3}{10}$	0.3
$\frac{4}{10}$	0.4
$\frac{5}{10}$	0.5
$\frac{6}{10}$	0.6
$\frac{7}{10}$	0.7
$\frac{8}{10}$	0.8
$\frac{9}{10}$	0.9

Directions: *From memory, convert the fractions to decimals.*

1) $\frac{3}{5} =$ 2) $\frac{1}{3} =$ 3) $\frac{4}{10} =$

4) $\frac{3}{4} =$ 5) $\frac{1}{2} =$ 6) $\frac{4}{5} =$

7) $\frac{7}{10} =$ 8) $\frac{1}{4} =$ 9) $\frac{1}{10} =$

10) $\frac{6}{10} =$ 11) $\frac{5}{10} =$ 12) $\frac{8}{10} =$

13) $\frac{2}{10} =$ 14) $\frac{1}{5} =$ 15) $\frac{2}{4} =$

16) $\frac{9}{10} =$ 17) $\frac{2}{3} =$ 18) $\frac{3}{10} =$

19) $\frac{2}{5} =$ 20) $\frac{1}{4} =$ 21) $\frac{1}{2} =$

Name: Date: Score: Unstoppable Owl

24) Fractions & Decimals (Common Conversions)

Common Conversions	
Fraction	**Decimal**
$\frac{1}{2}$	1)
$\frac{1}{3}$	2)
$\frac{2}{3}$	3)
$\frac{1}{4}$	4)
$\frac{2}{4}$	5)
$\frac{3}{4}$	6)
$\frac{1}{5}$	7)
$\frac{2}{5}$	8)
$\frac{3}{5}$	9)
$\frac{4}{5}$	10)
$\frac{1}{10}$	11)
$\frac{2}{10}$	12)
$\frac{3}{10}$	13)
$\frac{4}{10}$	14)
$\frac{5}{10}$	15)
$\frac{6}{10}$	16)
$\frac{7}{10}$	17)
$\frac{8}{10}$	18)
$\frac{9}{10}$	19)

Directions: *From memory, convert the chart's fractions to decimals. Then, cover the chart and convert the decimals into fractions by writing the numerator.*

20) $0.25 = \frac{\quad}{4}$

21) $0.7 = \frac{\quad}{10}$

22) $0.8 = \frac{\quad}{5}$

23) $0.2 = \frac{\quad}{10}$

24) $0.9 = \frac{\quad}{10}$

25) $0.\overline{3} = \frac{\quad}{3}$

26) $0.4 = \frac{\quad}{5}$

27) $0.5 = \frac{\quad}{10}$

28) $0.1 = \frac{\quad}{10}$

29) $0.6 = \frac{\quad}{5}$

30) $0.75 = \frac{\quad}{4}$

31) $0.4 = \frac{\quad}{10}$

32) $0.5 = \frac{\quad}{2}$

33) $0.\overline{6} = \frac{\quad}{3}$

34) $0.3 = \frac{\quad}{10}$

35) $0.5 = \frac{\quad}{4}$

36) $0.8 = \frac{\quad}{10}$

37) $0.2 = \frac{\quad}{5}$

38) $0.6 = \frac{\quad}{10}$

39) $0.75 = \frac{\quad}{4}$

40) $0.\overline{3} = \frac{\quad}{3}$

Name: Date: Score: Unstoppable Owl

25) Fractions & Decimals (Common Conversions)

Other fractions may be mentally converted to decimals too. A little multiplication can often help with the conversions. For these fractions, you can multiply and then write the product as hundredths when creating the decimal.

For eleven twentieths, 11 multiplied by 5 is 55. So, 55 hundredths is the decimal, and it is written as 0.55.

When the **denominator** is...	**Multiply** the **numerator** by...
20	5
25	4
50	2
100	1

Common Conversions

Fraction	Decimal
$\frac{1}{20}$	0.05
$\frac{2}{20}$	0.1
$\frac{5}{20}$	0.25
$\frac{11}{20}$	0.55
$\frac{19}{20}$	0.95
$\frac{1}{25}$	0.04
$\frac{2}{25}$	0.08
$\frac{6}{25}$	0.24
$\frac{12}{25}$	0.48
$\frac{21}{25}$	0.84
$\frac{1}{50}$	0.02
$\frac{2}{50}$	0.04
$\frac{8}{50}$	0.16
$\frac{23}{50}$	0.46
$\frac{41}{50}$	0.82
$\frac{1}{100}$	0.01
$\frac{2}{100}$	0.02
$\frac{28}{100}$	0.28
$\frac{73}{100}$	0.73

Directions: *Mentally convert the fractions to decimals.*

1) $\frac{4}{20} =$

2) $\frac{3}{25} =$

3) $\frac{6}{50} =$

4) $\frac{19}{100} =$

5) $\frac{9}{20} =$

6) $\frac{10}{25} =$

7) $\frac{14}{50} =$

8) $\frac{20}{25} =$

9) $\frac{7}{100} =$

10) $\frac{10}{20} =$

11) $\frac{18}{20} =$

12) $\frac{11}{25} =$

13) $\frac{34}{50} =$

14) $\frac{57}{100} =$

15) $\frac{3}{50} =$

16) $\frac{22}{25} =$

17) $\frac{15}{20} =$

18) $\frac{13}{25} =$

19) $\frac{87}{100} =$

20) $\frac{5}{25} =$

21) $\frac{18}{50} =$

Name: Date: Score: Unstoppable Owl

26) Fractions & Decimals (Common Conversions)

Common Conversions	
Fraction	**Decimal**
$\frac{3}{20}$	1)
$\frac{4}{20}$	2)
$\frac{8}{20}$	3)
$\frac{14}{20}$	4)
$\frac{17}{20}$	5)
$\frac{4}{25}$	6)
$\frac{9}{25}$	7)
$\frac{14}{25}$	8)
$\frac{17}{25}$	9)
$\frac{24}{25}$	10)
$\frac{4}{50}$	11)
$\frac{7}{50}$	12)
$\frac{12}{50}$	13)
$\frac{25}{50}$	14)
$\frac{49}{50}$	15)
$\frac{6}{100}$	16)
$\frac{8}{100}$	17)
$\frac{33}{100}$	18)
$\frac{81}{100}$	19)

Directions: *Mentally convert the chart's fractions to decimals. Then, mentally use division to convert the decimals into fractions and write the missing numerator.*

20) $0.42 = \frac{\quad}{50}$

21) $0.97 = \frac{\quad}{100}$

22) $0.6 = \frac{\quad}{20}$

23) $0.8 = \frac{\quad}{25}$

24) $0.58 = \frac{\quad}{50}$

25) $0.35 = \frac{\quad}{20}$

26) $0.28 = \frac{\quad}{25}$

27) $0.63 = \frac{\quad}{100}$

28) $0.65 = \frac{\quad}{20}$

29) $0.44 = \frac{\quad}{50}$

30) $0.28 = \frac{\quad}{25}$

31) $0.18 = \frac{\quad}{50}$

32) $0.03 = \frac{\quad}{100}$

33) $0.3 = \frac{\quad}{20}$

34) $0.1 = \frac{\quad}{50}$

35) $0.32 = \frac{\quad}{25}$

36) $0.8 = \frac{\quad}{20}$

37) $0.34 = \frac{\quad}{50}$

38) $0.26 = \frac{\quad}{100}$

39) $0.92 = \frac{\quad}{25}$

40) $0.76 = \frac{\quad}{50}$

Name: Date: Score: Unstoppable Owl

27) Fractions & Decimals (Common Conversions)

Common Conversions	
Fraction	**Decimal**
$\frac{1}{2}$	1)
$\frac{1}{3}$	2)
$\frac{2}{3}$	3)
$\frac{1}{4}$	4)
$\frac{3}{4}$	5)
$\frac{3}{5}$	6)
$\frac{4}{5}$	7)
$\frac{1}{10}$	8)
$\frac{9}{10}$	9)
$\frac{1}{20}$	10)
$\frac{7}{20}$	11)
$\frac{2}{25}$	12)
$\frac{6}{25}$	13)
$\frac{12}{25}$	14)
$\frac{7}{50}$	15)
$\frac{23}{50}$	16)
$\frac{31}{50}$	17)
$\frac{29}{100}$	18)
$\frac{77}{100}$	19)

Mixed Review: *From memory, convert the chart's fractions to decimals. Then, cover the chart and convert the decimals into fractions by writing the numerator.*

20) $0.7 = \frac{}{10}$

21) $0.55 = \frac{}{20}$

22) $0.4 = \frac{}{5}$

23) $0.36 = \frac{}{25}$

24) $0.57 = \frac{}{100}$

25) $0.5 = \frac{}{4}$

26) $0.26 = \frac{}{50}$

27) $0.4 = \frac{}{10}$

28) $0.\overline{6} = \frac{}{3}$

29) $0.04 = \frac{}{25}$

30) $0.2 = \frac{}{5}$

31) $0.09 = \frac{}{100}$

32) $0.5 = \frac{}{2}$

33) $0.\overline{3} = \frac{}{3}$

34) $0.98 = \frac{}{50}$

35) $0.75 = \frac{}{20}$

36) $0.25 = \frac{}{4}$

37) $0.2 = \frac{}{10}$

38) $0.9 = \frac{}{20}$

39) $0.8 = \frac{}{50}$

40) $0.16 = \frac{}{25}$

Unstoppable Owl

Name: Date: Score:

28) Identifying Mixed Fractions

Fractions can be used to represent parts of a whole. They can be used to represent whole numbers too. **Mixed fractions** represent a whole number and a fraction. The number on the left side of the mixed fraction represents the whole number and the fraction to the right of it represents the parts of a whole.

Examples:

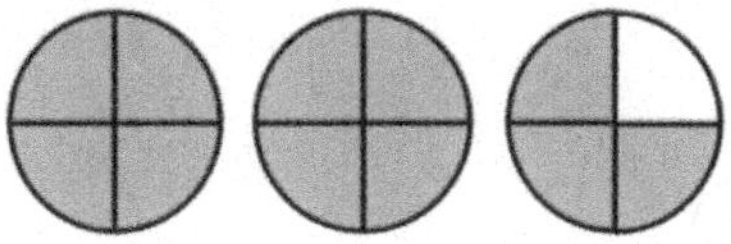 $= 2\frac{3}{4}$

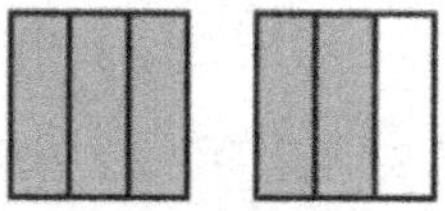 $= 1\frac{2}{3}$

Directions: *Write a mixed fraction for each set of illustrated fractions.*

1) 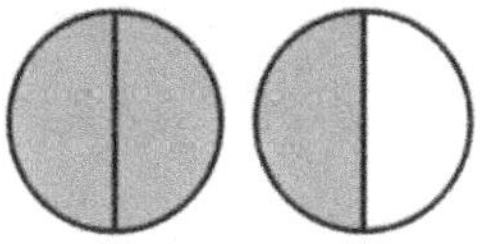 =

2) 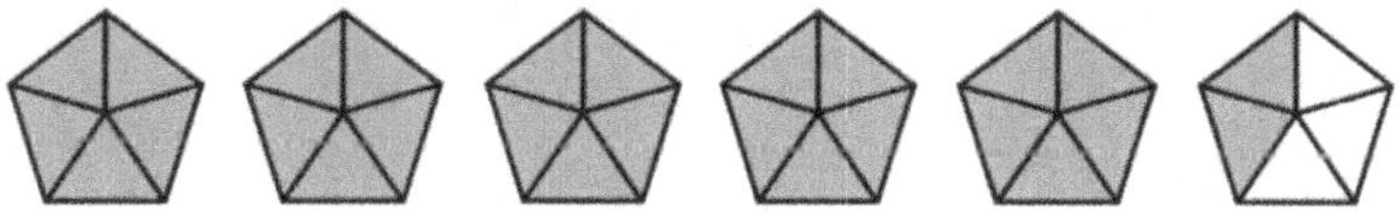 =

3)  =

4) 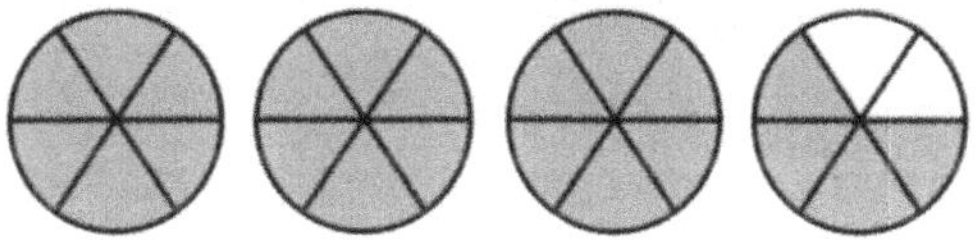 =

5) 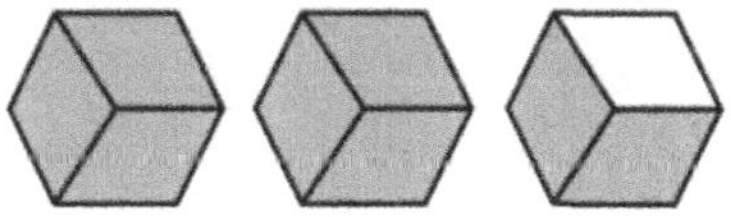 =

6)  =

7) 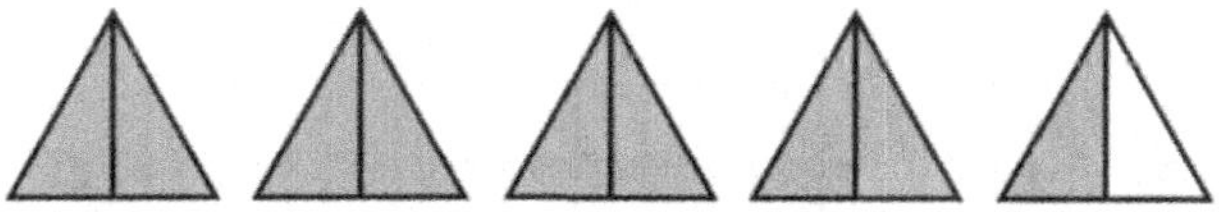 =

8) 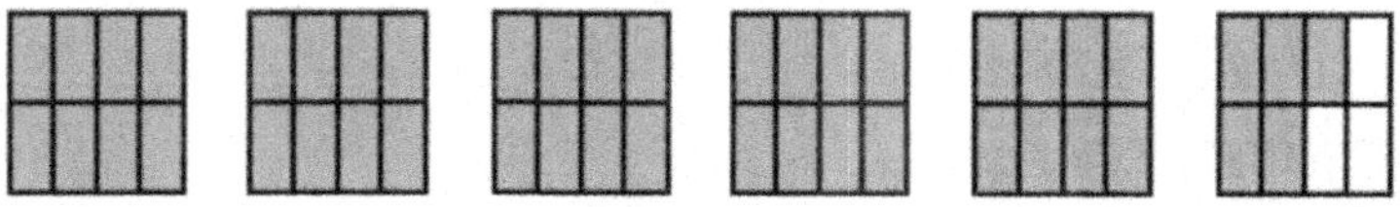 =

Name: Date: Score:

29) Identifying Mixed Fractions

Directions: *Write a mixed fraction for each set of illustrated fractions.*

1) 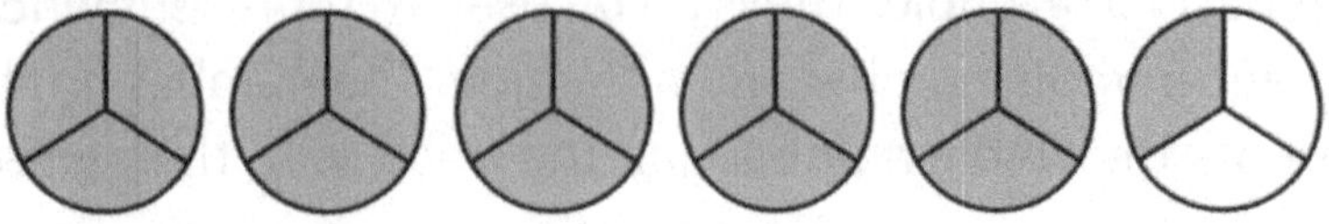 =

2) 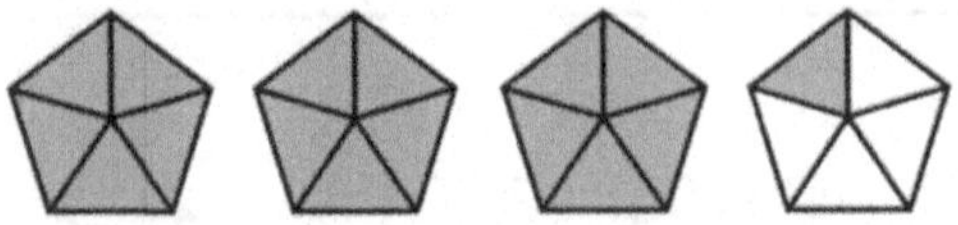 =

3) 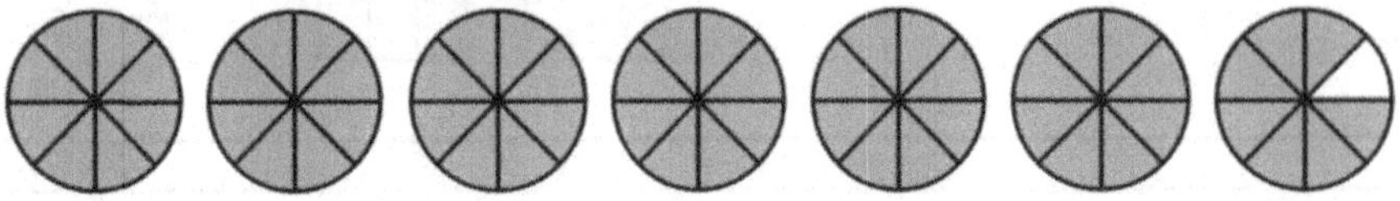 =

4) 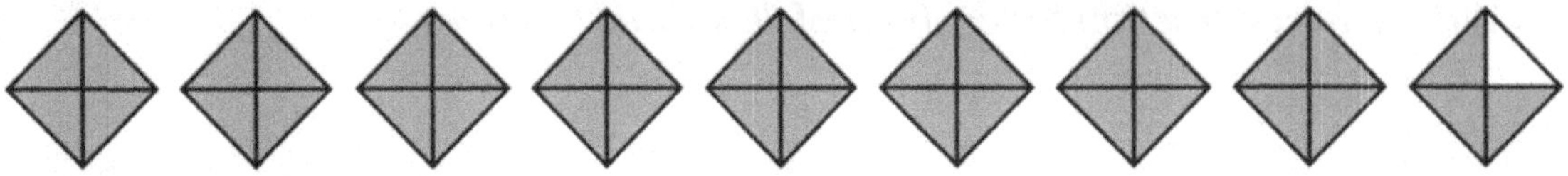 =

5) 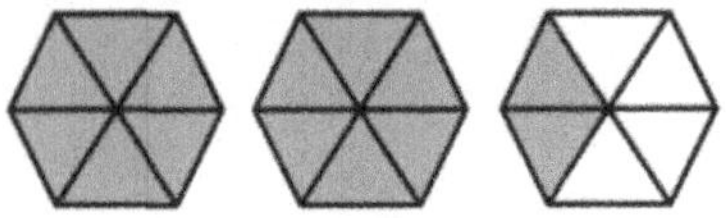 =

6)  =

7) 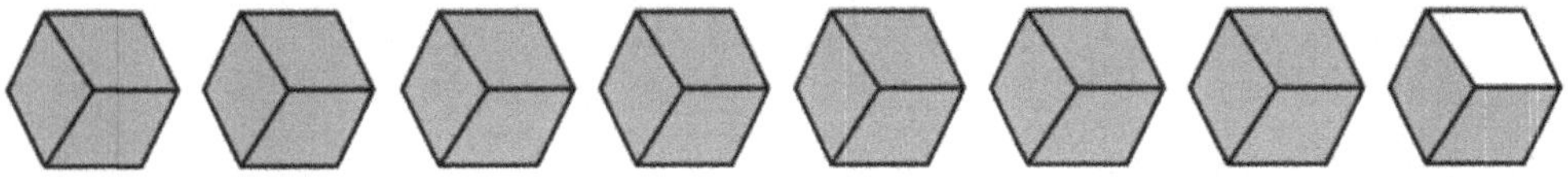 =

8) 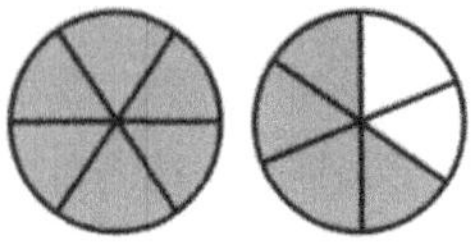 =

9) 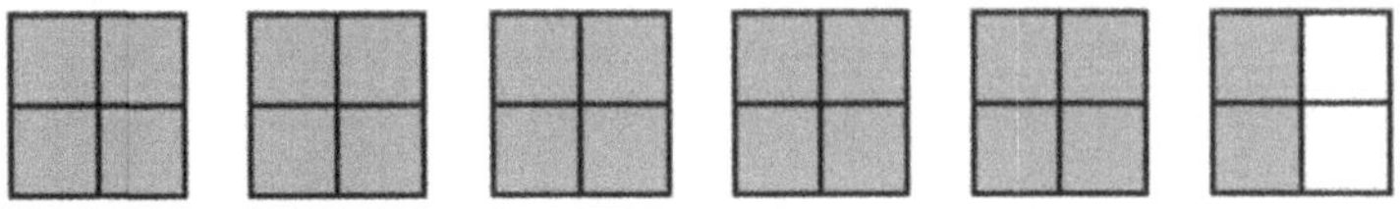 =

10)  =

11)  =

12) 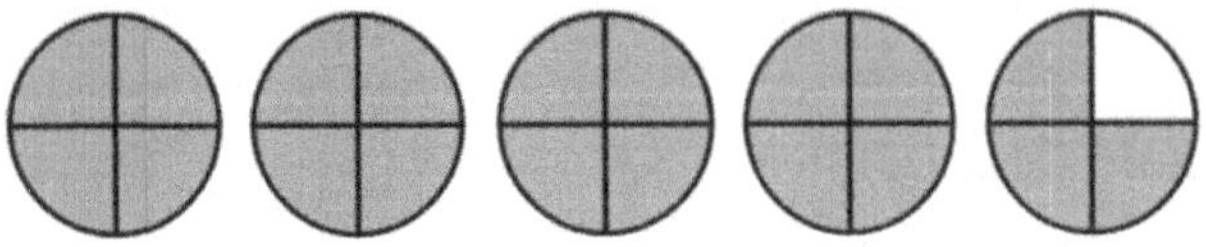 =

Name: Date: Score:

30) Identifying Mixed Fractions

Directions: *Shade the illustrations to represent each mixed fraction.*
(Some illustrations may not be needed.)

1) $4\frac{1}{4}$ = 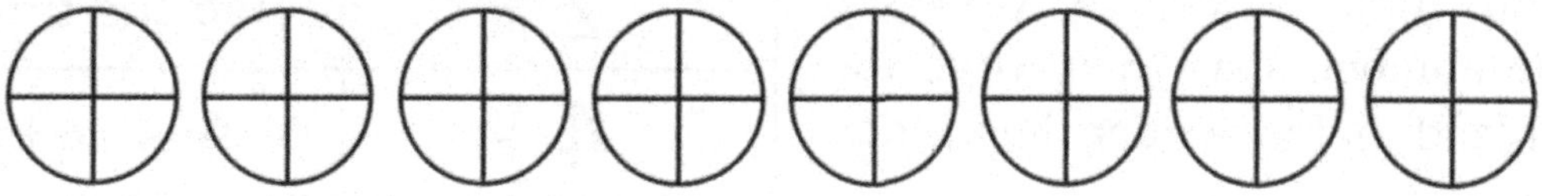

2) $7\frac{1}{2}$ = 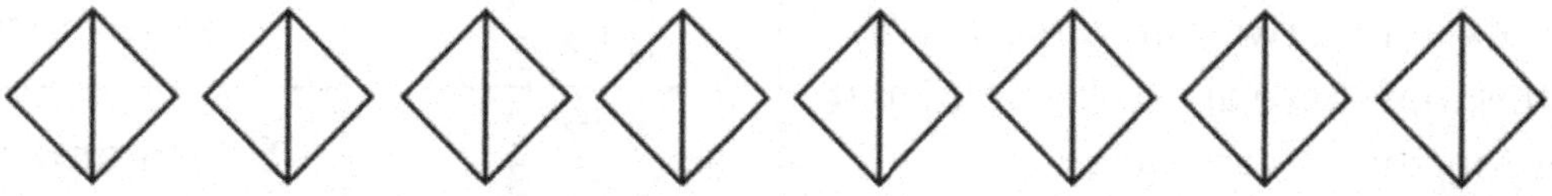

3) $1\frac{3}{5}$ = 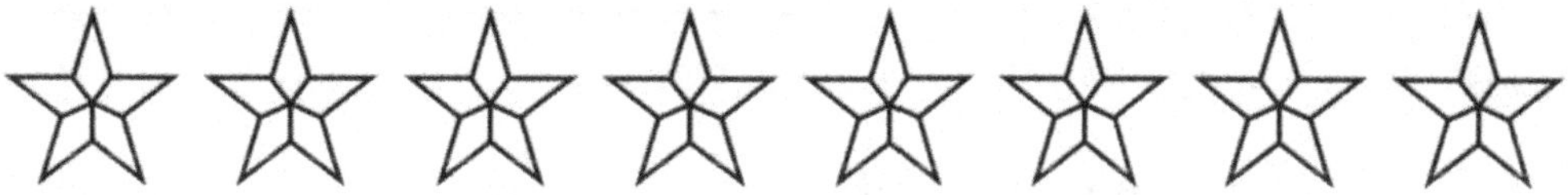

4) $5\frac{2}{3}$ = 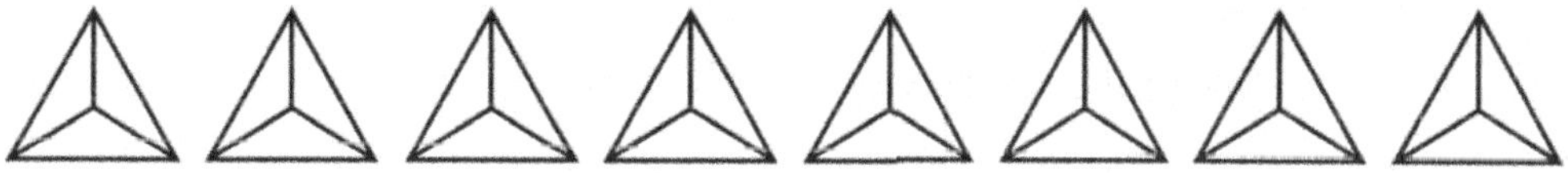

5) $2\frac{3}{8}$ = 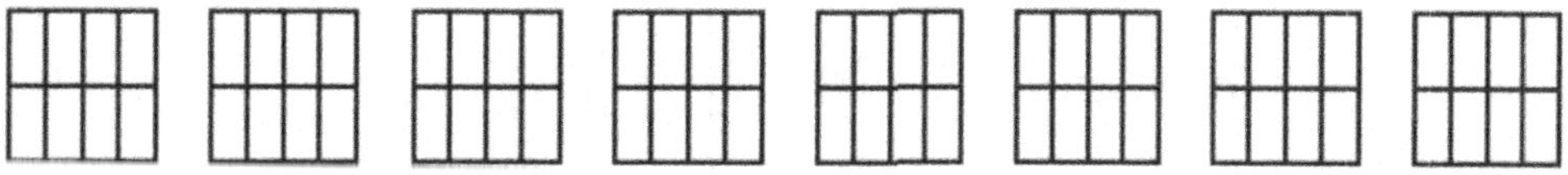

6) $6\frac{3}{4}$ = 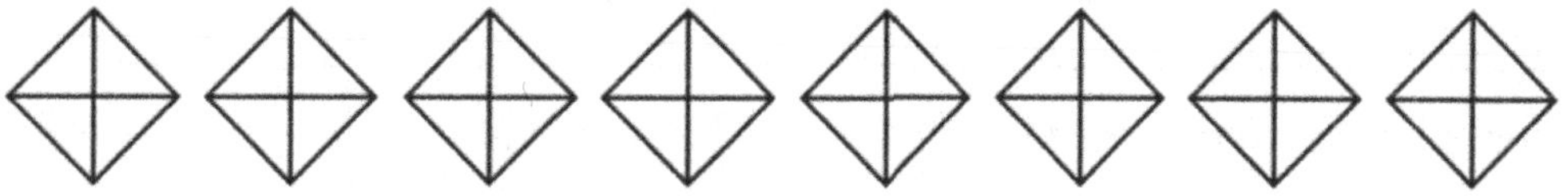

7) $3\frac{5}{6}$ = 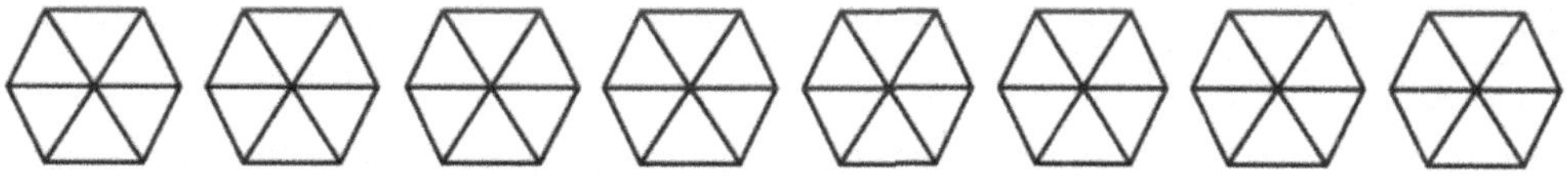

8) $5\frac{1}{2}$ = 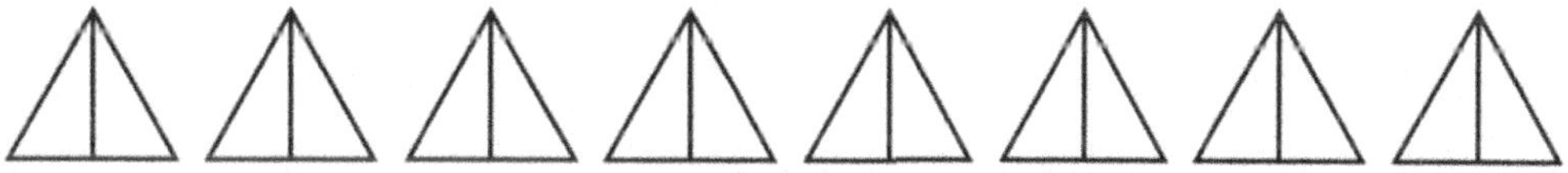

9) $7\frac{1}{3}$ = 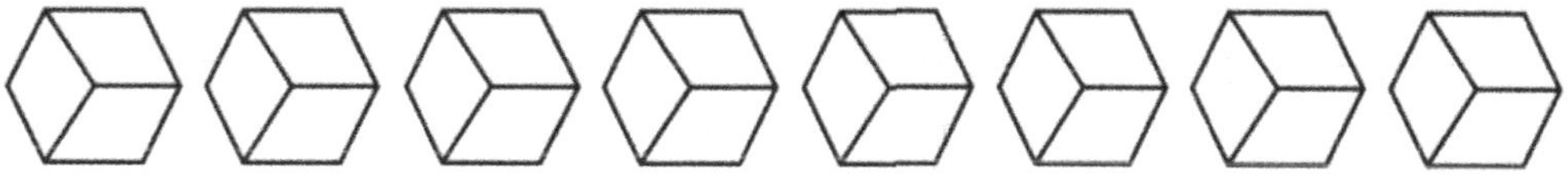

10) $4\frac{4}{5}$ = 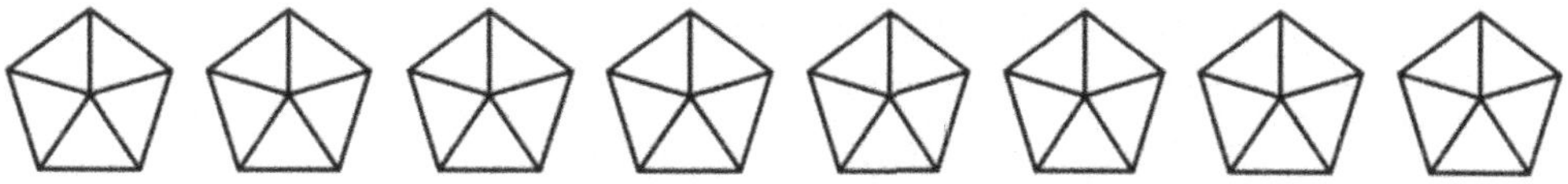

11) $6\frac{2}{4}$ = 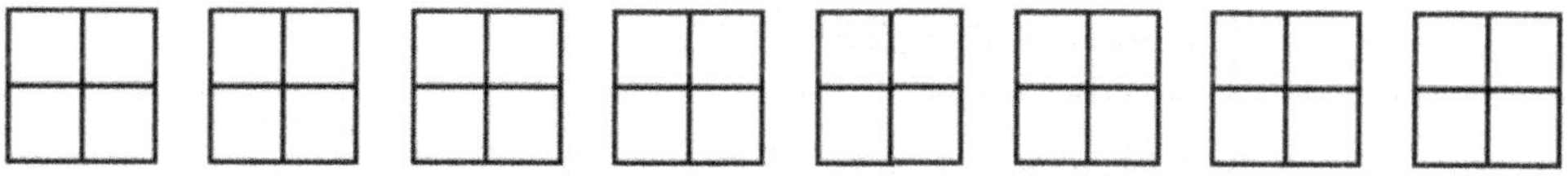

12) $2\frac{1}{5}$ = 

Name: Date: Score: Unstoppable Owl

31) Identifying Mixed Fractions

Mixed fractions can also be written in word form. To do this, write the whole number first. Then, write the word "and" before writing the fraction in word form.

For example, *two and three-fourths* signifies that there are two wholes and three-fourths of a whole.

Number Form:	Word Form:
$2\frac{3}{4}$	Two and three-fourths
$4\frac{1}{2}$	Four and one-half
$10\frac{4}{5}$	Ten and four-fifths
$7\frac{2}{3}$	Seven and two-thirds

Directions: *Draw a line to match the number form of each mixed fraction with its word form.*

$3\frac{2}{5}$	1)	A. **One and one-half**	$7\frac{1}{4}$	6)	A. **Two and one-twentieth**
$8\frac{1}{3}$	2)	B. **Six and four-sevenths**	$2\frac{5}{12}$	7)	B. **Two and five-twelfths**
$6\frac{4}{7}$	3)	C. **Nine and two-fourths**	$20\frac{2}{5}$	8)	C. **Seven and one-fourth**
$1\frac{1}{2}$	4)	D. **Three and two-fifths**	$12\frac{6}{10}$	9)	D. **Twenty and two-fifths**
$9\frac{2}{4}$	5)	E. **Eight and one-third**	$2\frac{1}{20}$	10)	E. **Twelve and six-tenths**

Directions: *Write each word form in number form and write each number form in word form.*

Number Form:	Word Form:
$6\frac{5}{8}$	11)
$3\frac{7}{9}$	12)
$15\frac{41}{100}$	13)
$4\frac{1}{2}$	14)
$33\frac{3}{5}$	15)
16)	One and one-eleventh
17)	Thirteen and three-eighths
18)	Seven and eight-ninths
19)	Thirty and thirty-three hundredths
20)	Three and seven-thirteenths

Name: Date: Score: Unstoppable Owl

32) Identifying Mixed Fractions

Directions: *Write a mixed fraction for each set of illustrated fractions.*

1) 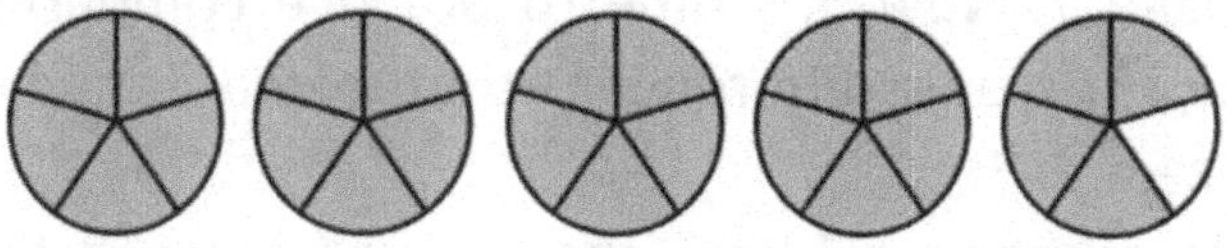 =

2)  =

3) 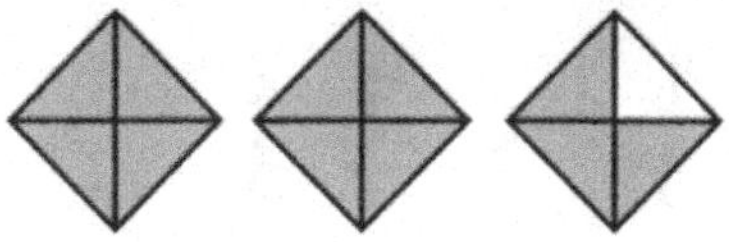 =

4) 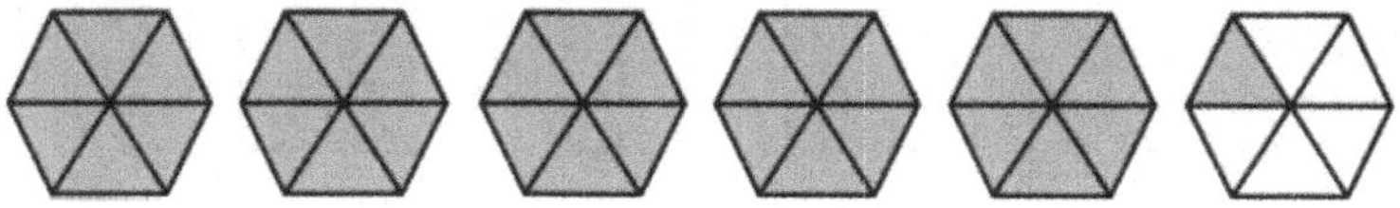 =

5)  =

Directions: *Write each word form in number form and write each number form in word form.*

Number Form:	**Word Form:**
$5\frac{5}{9}$	6)
$100\frac{1}{3}$	7)
$16\frac{3}{11}$	8)
$2\frac{4}{7}$	9)
$11\frac{1}{6}$	10)
$6\frac{3}{4}$	11)
12)	**Four and sixteen twenty-fifths**
13)	**Twenty-four and two-fifths**
14)	**Eighteen and one-third**
15)	**Nine and ten-elevenths**
16)	**Thirty-seven and seven-eighths**
17)	**Ninety-nine and one-fourth**

Name: Date: Score: *Unstoppable Owl*

33) Comparing Mixed Fractions (With Like Denominators)

Comparing fractions was introduced on page 7. Now, it's time to practice comparing mixed fractions. Many of the same skills apply. The only difference is that mixed fractions have whole numbers that need to be considered.

The mixed fraction with the greater whole number will have the greater value. If the whole numbers are the same, the fraction portions of the mixed fractions need to be compared to see which is greater.

Example 1: (Different Whole Numbers)

$$2\frac{1}{4} > 1\frac{3}{4}$$

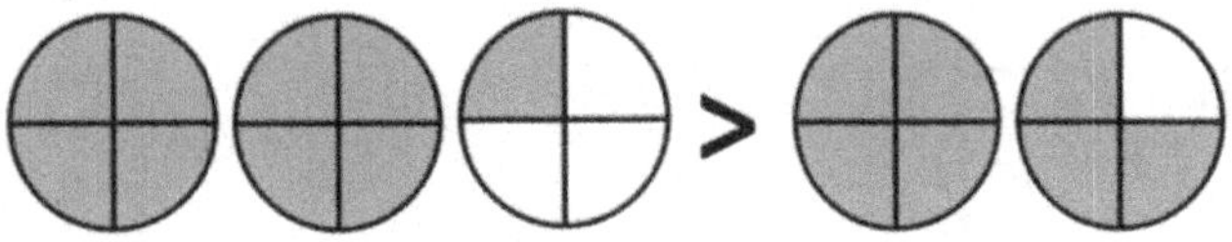

Two and one-fourth **is greater than** one and three-fourths.

Example 2: (Same Whole Numbers)

$$1\frac{1}{3} < 1\frac{2}{3}$$

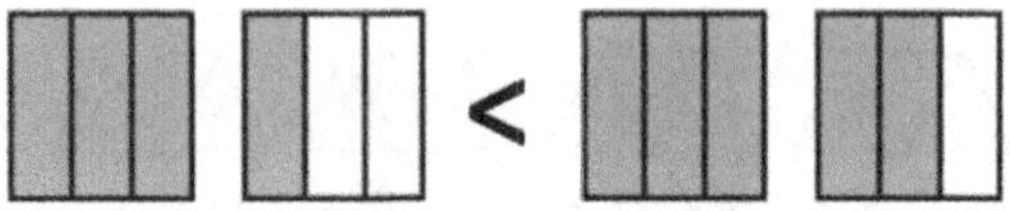

One and one-third **is less than** one and two-thirds.

Directions: *Draw a greater than or less than symbol to compare each set of fractions.*

1) 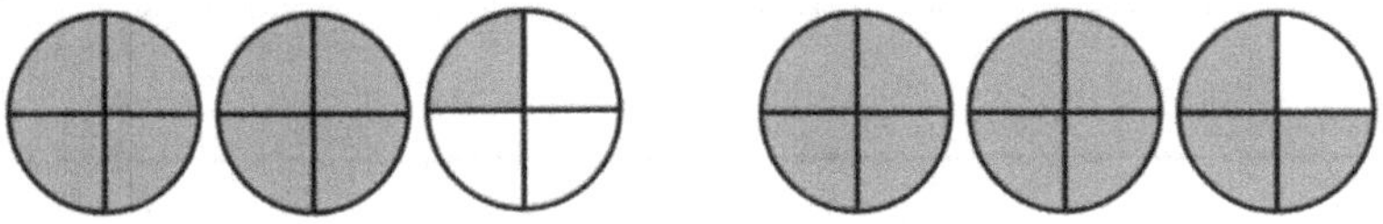

2) 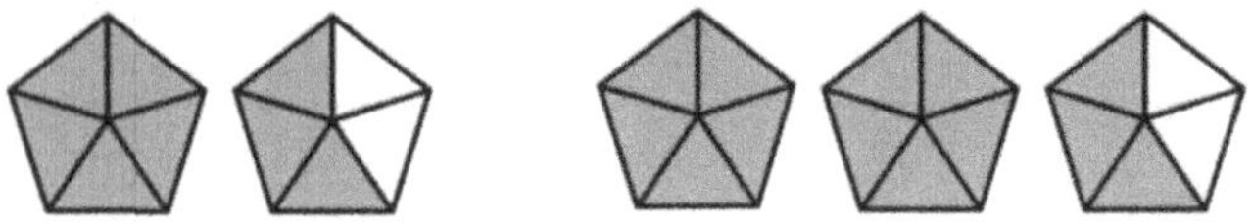

3) 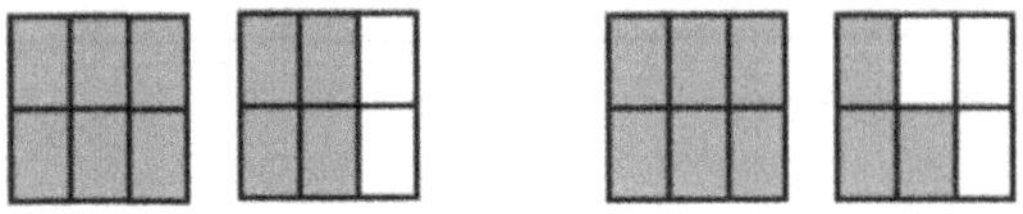

4) 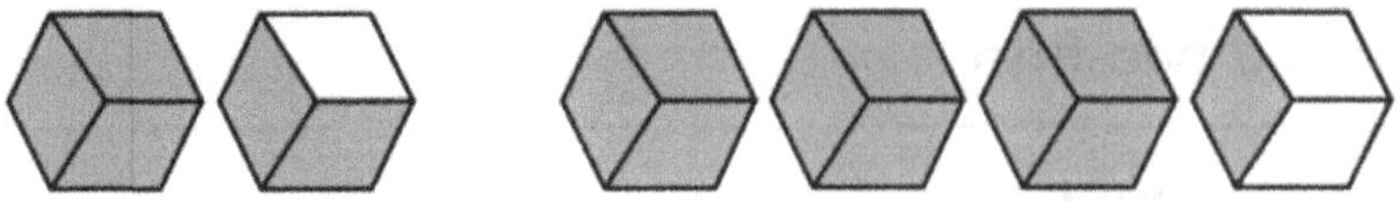

5) 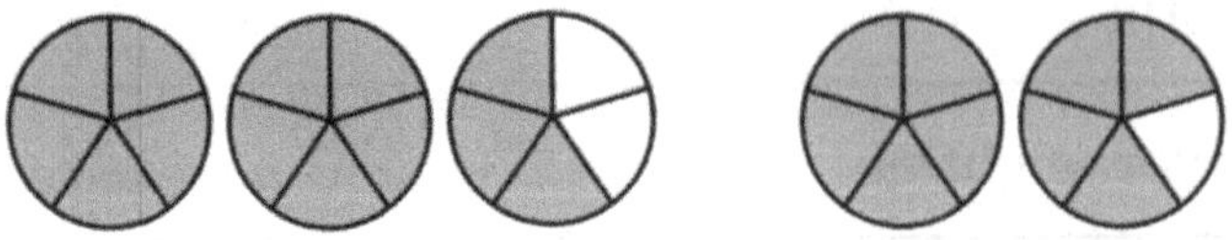

6) $4\frac{5}{7}$ $7\frac{3}{7}$

7) $6\frac{3}{4}$ $6\frac{2}{4}$

8) $7\frac{2}{5}$ $9\frac{4}{5}$

9) $8\frac{4}{9}$ $3\frac{8}{9}$

10) $4\frac{2}{6}$ $4\frac{3}{6}$

Name: Date: Score: Unstoppable Owl

34) Comparing Mixed Fractions (With Unlike Denominators)

Now, it's time to compare mixed fractions that have unlike denominators. You already learned all the skills to do this (Refer to the previous page and page 8.)

Comparing fractions with unlike denominators requires a little more thinking. Take your time and study each problem carefully. Compare the whole numbers first. If the same, compare the fractions.

Example 1:

$1\frac{2}{3} < 2\frac{1}{4}$

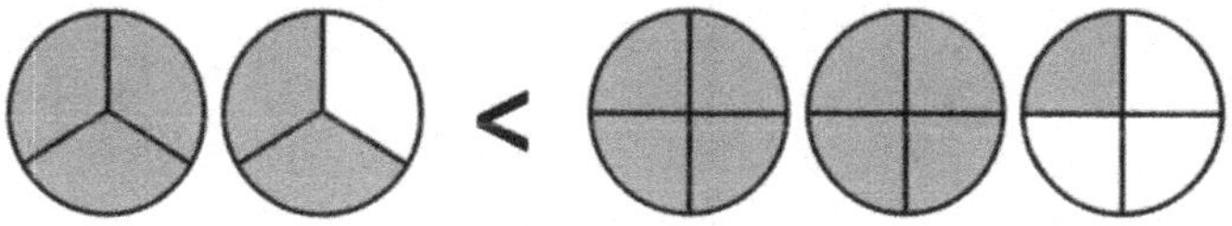

One and two-thirds **is less than** two and one-fourth.

Example 2:

$1\frac{1}{3} > 1\frac{1}{6}$

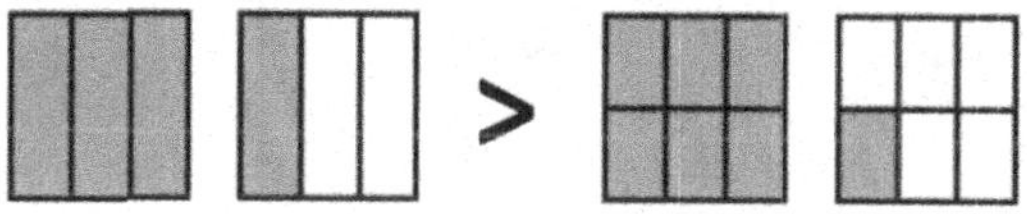

One and one-third **is greater than** one and one-sixth.

Directions: *Draw a greater than or less than symbol to compare each set of fractions.*

1) 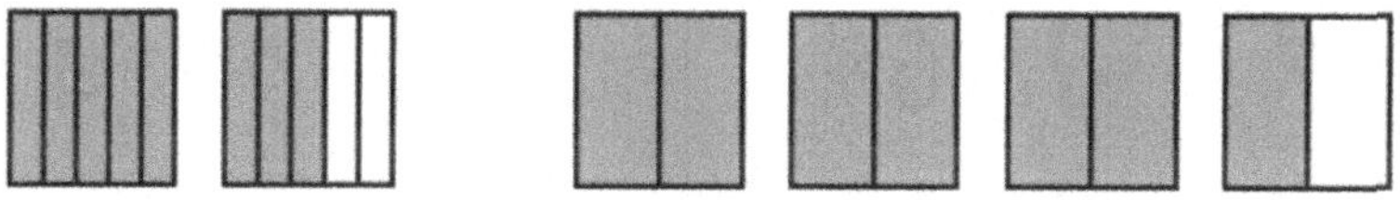

2) 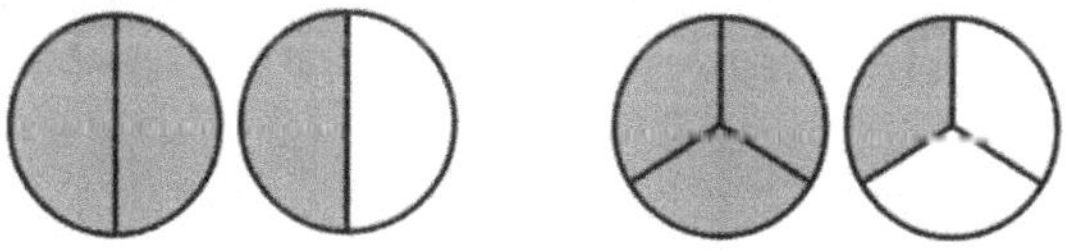

3) 

4) 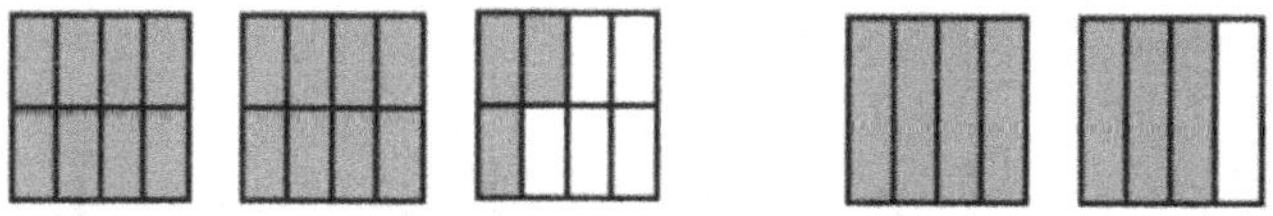

5) 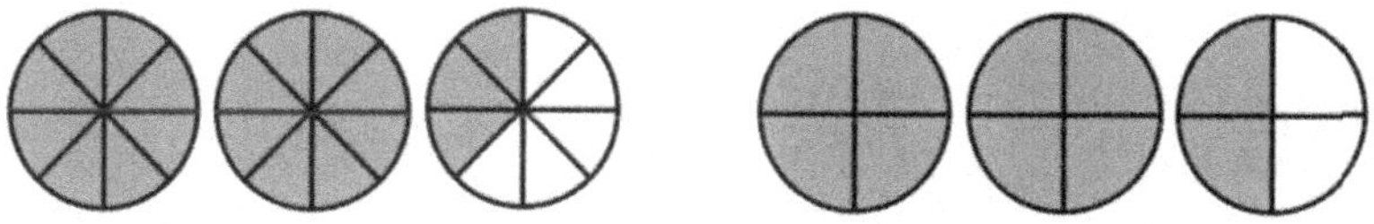

6) $5\frac{1}{4}$ $4\frac{3}{6}$

7) $2\frac{2}{3}$ $3\frac{2}{4}$

8) $9\frac{4}{8}$ $9\frac{3}{5}$

9) $5\frac{1}{4}$ $5\frac{1}{3}$

10) $8\frac{2}{6}$ $8\frac{3}{4}$

Name: Date: Score: Unstoppable Owl

35) Comparing Mixed Fractions (With Unlike Denominators)

This is the last page dedicated to only comparing mixed fractions. Focus on one column at a time. Take breaks if you need to. Make sure that you understand this skill before proceeding to the next page.

Directions: *Draw a greater than or less than symbol to compare each set of mixed fractions.*

1) $3\frac{9}{10}$ $8\frac{1}{2}$

2) $7\frac{3}{4}$ $7\frac{2}{7}$

3) $2\frac{4}{5}$ $1\frac{5}{10}$

4) $5\frac{1}{2}$ $4\frac{2}{3}$

5) $2\frac{6}{10}$ $4\frac{3}{10}$

6) $7\frac{1}{4}$ $5\frac{1}{2}$

7) $9\frac{2}{5}$ $9\frac{2}{3}$

8) $3\frac{4}{5}$ $3\frac{3}{4}$

9) $4\frac{2}{9}$ $6\frac{5}{6}$

10) $3\frac{7}{10}$ $4\frac{3}{5}$

11) $2\frac{35}{100}$ $2\frac{1}{3}$

12) $8\frac{1}{6}$ $6\frac{2}{6}$

13) $4\frac{4}{5}$ $4\frac{3}{4}$

14) $3\frac{1}{4}$ $1\frac{3}{5}$

15) $5\frac{1}{2}$ $6\frac{7}{9}$

16) $8\frac{3}{5}$ $8\frac{2}{3}$

17) $5\frac{2}{4}$ $4\frac{2}{3}$

18) $1\frac{3}{10}$ $2\frac{1}{5}$

19) $9\frac{2}{3}$ $9\frac{3}{4}$

20) $6\frac{8}{10}$ $7\frac{9}{20}$

21) $4\frac{1}{4}$ $3\frac{1}{2}$

22) $5\frac{1}{2}$ $5\frac{2}{5}$

23) $8\frac{6}{10}$ $9\frac{7}{100}$

24) $6\frac{2}{5}$ $2\frac{3}{8}$

25) $4\frac{1}{4}$ $4\frac{23}{100}$

26) $7\frac{7}{8}$ $5\frac{4}{5}$

27) $3\frac{2}{9}$ $2\frac{3}{6}$

28) $6\frac{2}{3}$ $6\frac{7}{10}$

29) $1\frac{1}{2}$ $1\frac{4}{10}$

30) $6\frac{3}{5}$ $3\frac{2}{3}$

Name: Date: Score: Unstoppable Owl

36) Mixed Fractions & Decimals (Conversions)

Like other fractions, mixed fractions can be represented by a decimal. A mixed fraction and decimal are simply different ways to write and express the value of a number.

You've practiced writing decimals for values that are less than 1. Now, you will practice writing decimals for values that include whole numbers. The **whole numbers** are written on the *left* side of the decimal. The **parts of a whole** are written on the *right* side of the decimal.

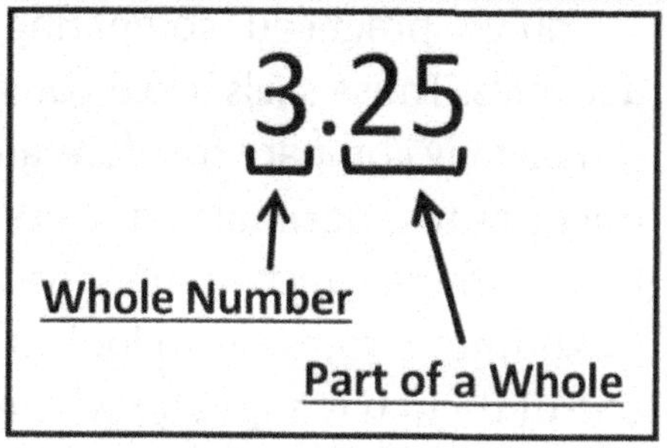

Examples:

$3\frac{1}{4} = 3.25$	$12\frac{3}{5} = 12.6$
The whole number is **3**. *One-fourth written as a decimal* is **0.25**. So, three and one-fourth is **3.25**.	*The whole number is* **12**. *Three-fifths written as a decimal* is **0.6**. So, twelve and three-fifths is **12.6**.

Directions: *Write each mixed fraction in decimal form, rounded to the* nearest hundredth.

1) $5\frac{3}{4} =$

2) $2\frac{4}{5} =$

3) $9\frac{7}{10} =$

4) $10\frac{3}{4} =$

5) $1\frac{2}{3} =$

6) $6\frac{1}{2} =$

7) $4\frac{2}{5} =$

8) $7\frac{1}{10} =$

9) $3\frac{1}{20} =$

10) $11\frac{1}{4} =$

11) $5\frac{9}{10} =$

12) $24\frac{1}{5} =$

13) $9\frac{3}{100} =$

14) $15\frac{3}{4} =$

15) $32\frac{2}{5} =$

16) $1\frac{3}{20} =$

17) $7\frac{1}{3} =$

18) $3\frac{1}{2} =$

19) $6\frac{3}{50} =$

20) $95\frac{4}{5} =$

21) $47\frac{2}{3} =$

Name: Date: Score: Unstoppable Owl

37) Comparing Fractions & Decimals

You've practiced comparing fractions. You've also practiced converting fractions to decimals. These skills allow you to compare fractions and decimals.

You may compare fractions and decimals in your mind, but it is often helpful to convert any fractions to a decimal first. Once all the values are written in decimal form, it is usually easier to determine which number has the greatest value (or if they have an equal value).

Another strategy is to look at the whole numbers for the decimals and mixed fractions. If one of them has a greater whole number than the other, then it has the greater value.

Example: (Converting fractions to decimals)	**Example:** (Comparing the whole numbers)
$5\frac{3}{4} > 5.6$ *(5.75)*	$2\frac{3}{5} < 4.5$
The whole numbers are the same. Five and three-fourths written as a decimal is 5.75, which is greater than 5.6.	*The whole numbers are 2 and 4. 4 is greater than 2.* Therefore, 4.5 is greater than two and three fifths.

Directions: *Draw a <, >, or = symbol to compare the fractions and decimals.*

1) $4\frac{4}{5}$ 2.75	8) 6.5 $6\frac{1}{2}$	15) $7\frac{9}{10}$ 7.88
2) 8.2 $8\frac{1}{4}$	9) $5\frac{3}{20}$ 5.1	16) 4.3 $5\frac{1}{4}$
3) 5.5 $5\frac{2}{5}$	10) $8\frac{2}{5}$ 6.4	17) 2.15 $2\frac{4}{25}$
4) $1\frac{7}{10}$ 2.4	11) 1.6 $1\frac{2}{3}$	18) $\frac{2}{3}$ 0.6
5) $7\frac{3}{5}$ 7.6	12) $\frac{3}{5}$ 0.7	19) 3.08 $4\frac{1}{10}$
6) 9.2 $5\frac{4}{5}$	13) $9\frac{3}{10}$ 10.2	20) $10\frac{7}{20}$ 10.35
7) $3\frac{1}{4}$ 3.28	14) 0.25 $\frac{1}{3}$	21) 6.8 $6\frac{3}{4}$

Name: Date: Score: Unstoppable Owl

38) Comparing Fractions & Decimals

Compare each set of numbers and then order them from least to greatest. *Remember that it may be helpful to first write the value of each number in decimal form.*

Directions: *Compare each set of numbers then write them in order from least to greatest.*

1) $2\frac{3}{4}$, 1.5, $1\frac{1}{4}$, 2.65

2) $5\frac{1}{2}$, 5.4, 5.2, $5\frac{1}{4}$

3) 3.75, $4\frac{3}{5}$, 4.5, $3\frac{4}{5}$

4) 8.2, $7\frac{9}{10}$, $7\frac{4}{5}$, 7.85

5) $6\frac{3}{4}$, $6\frac{6}{10}$, 6.65, $6\frac{2}{3}$

6) 0.8, $\frac{7}{10}$, 1.5, $2\frac{3}{5}$

7) 9.25, $9\frac{1}{5}$, 10.2, 9.4

8) $3\frac{3}{10}$, 3.35, $3\frac{1}{3}$, 3.32

9) $6\frac{1}{2}$, 7.4, 6.6, $5\frac{3}{5}$

10) 2.44, $2\frac{2}{5}$, 2.5, $2\frac{39}{100}$

11) 11.4, $10\frac{9}{20}$, $11\frac{1}{2}$, 10.41

12) $5\frac{2}{3}$, $6\frac{7}{10}$, 5.55, $6\frac{3}{4}$

13) $7\frac{2}{7}$, 7.5, 6.8, $7\frac{2}{9}$

14) 2.2, $2\frac{1}{10}$, $2\frac{1}{4}$, 2.05

15) $4\frac{1}{5}$, 5.4, $4\frac{7}{20}$, 4.33

Name: Date: Score:

39) Improper Fractions

You've practiced writing mixed fractions, which consist of a whole number and a fraction. **Improper fractions** are similar to mixed fractions. They represent numbers that have one or more whole parts, and often include parts of a whole too. The difference with improper fractions is that they do not combine a whole number and a fraction. Improper fractions are written as a fraction that has a numerator that is equal to or greater than the denominator.

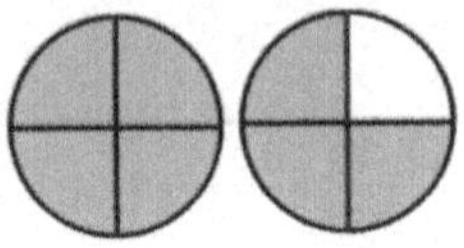

$= 1\frac{3}{4} = \frac{7}{4}$

Mixed Fraction

Improper Fraction

With an improper fraction, the numerator is the total number of parts, and the denominator is the number of parts that are in each whole. In the example to the left, there are 7 total parts shaded and 4 equal parts in each whole. So, this improper fraction is seven-fourths.

Examples:

$= \frac{12}{5}$

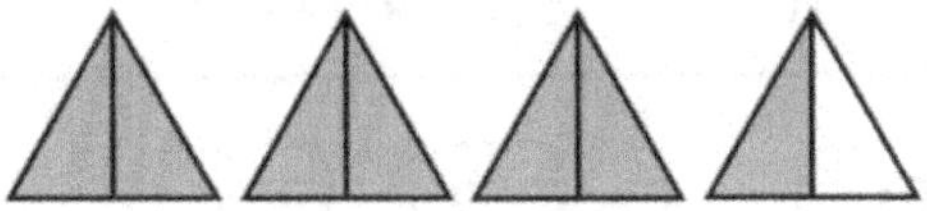

$= \frac{7}{2}$

Directions: *Write an improper fraction for each set of illustrations.*

1) 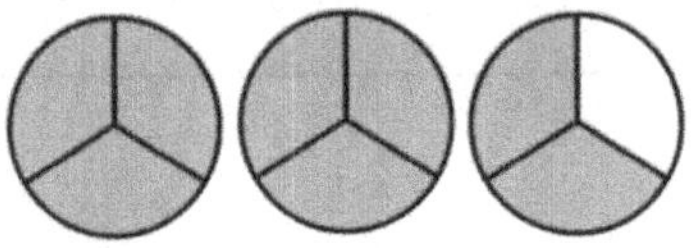 =

2) 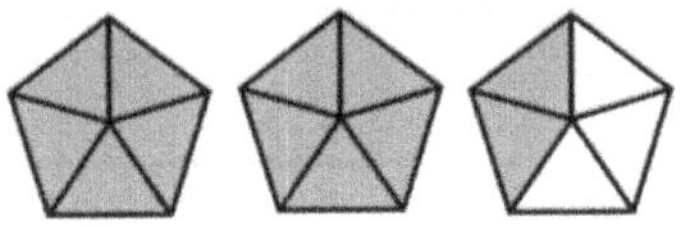 =

3) 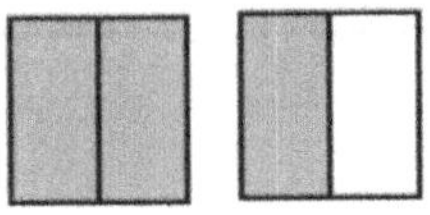 =

4) 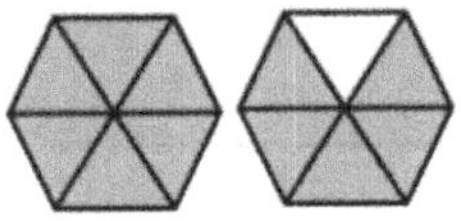 =

5) 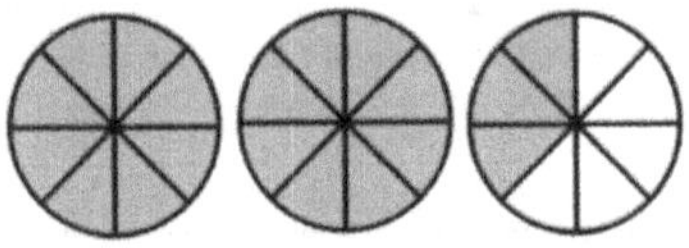 =

6) 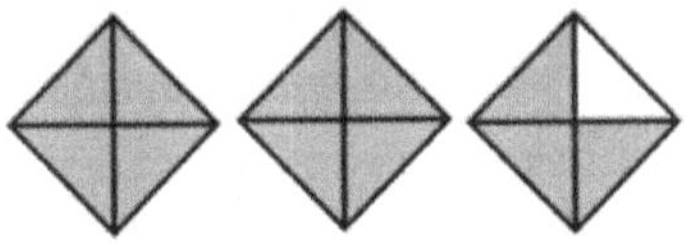 =

7) 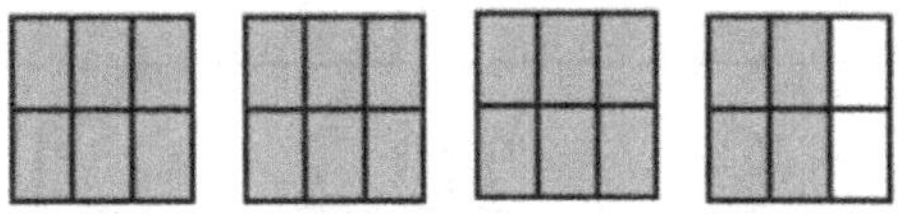 =

8)  =

9) 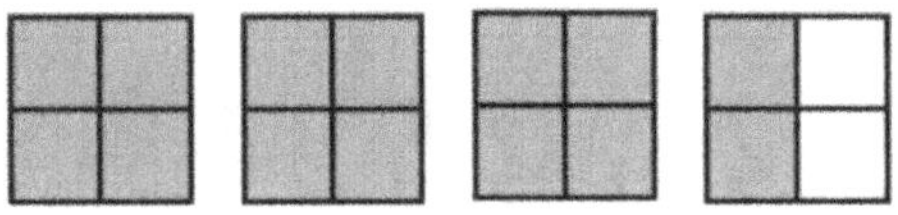 =

10) 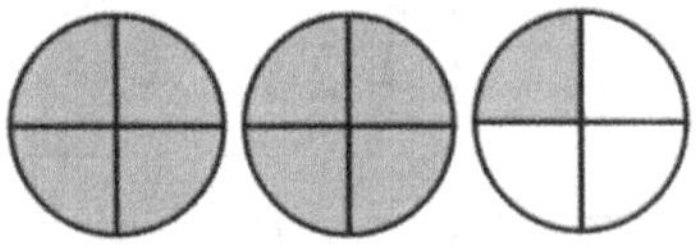 =

11) 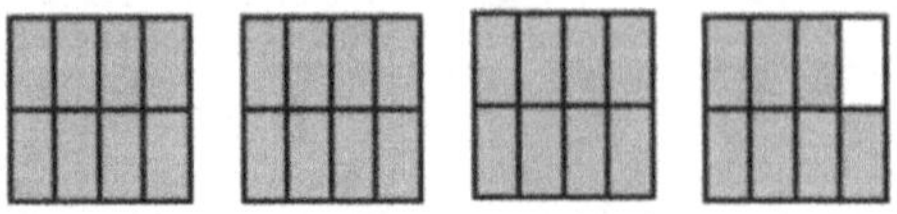 =

12) 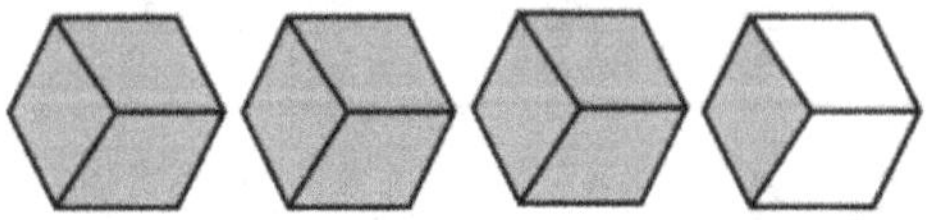 =

Name: Date: Score:

40) Improper Fractions

Improper fractions can be used to represent a whole number too. For instance, the first example below shows 3 wholes. Since each of these wholes are divided into 6 pieces, the irrational fraction for these 3 wholes is eighteen-sixths.

Have you noticed that multiplication can be used to help calculate an irrational fraction? The number of wholes multiplied by the number of parts in each whole is equal to the number of parts in those wholes. This can then be added to any remaining parts of a whole that are left. Try multiplying to calculate irrational numbers. It may save you time.

Example 1:

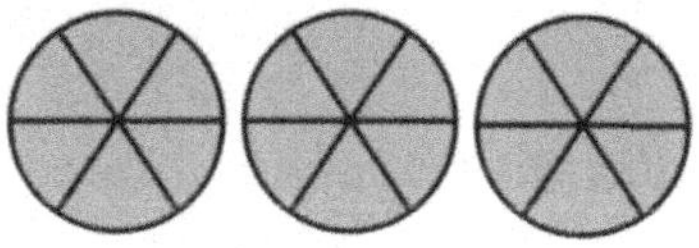

$= \frac{18}{6}$

There are **3** *wholes.*
There are **6** *parts in each whole.*
3 × 6 = **18**.
Are there any parts of a whole? **No**

Example 2:

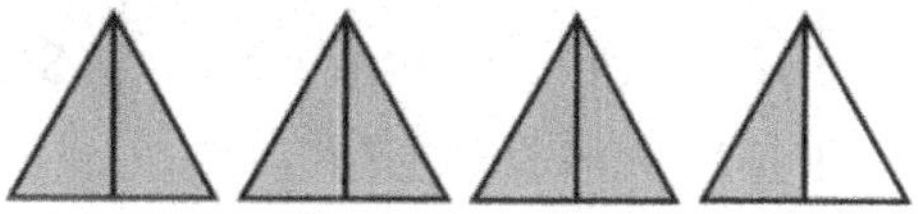

$= \frac{7}{2}$

There are **3** *wholes.*
There are **2** *parts in each whole.*
3 × 2 = **6**.
Are there any parts of a whole? **Yes, 1 part.**
6 + 1 = **7**.

Directions: *Write an improper fraction for each set of illustrations.*

1) 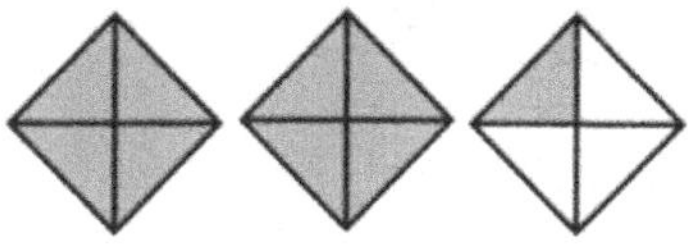 =

2) 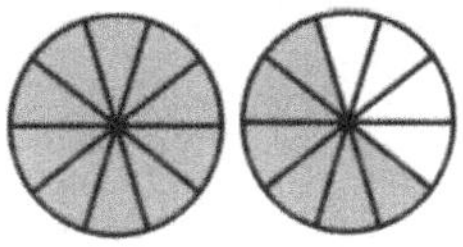 =

3) 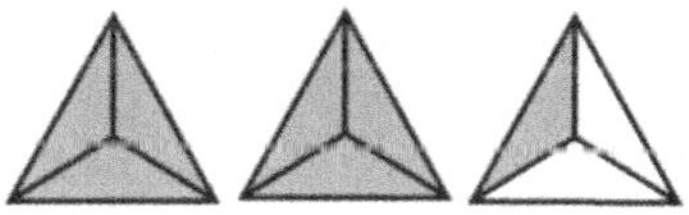 =

4)  =

5) 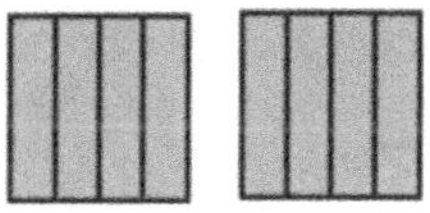 =

6) 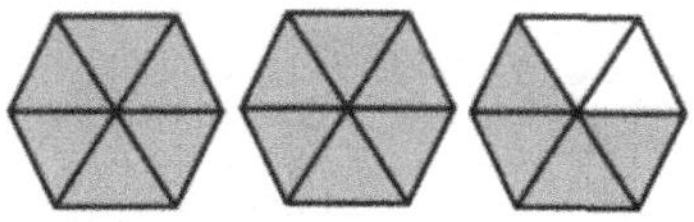 =

7) 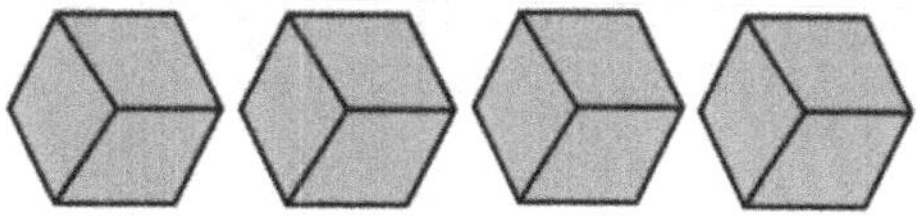 =

8) 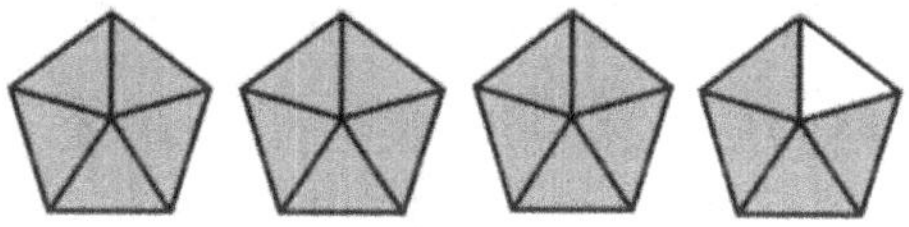 =

9) 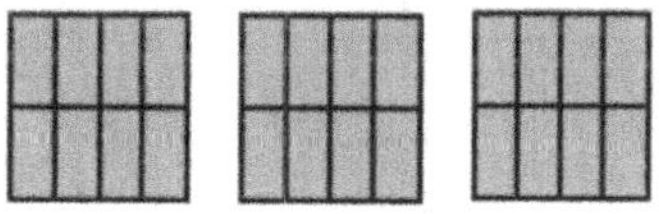 =

10) 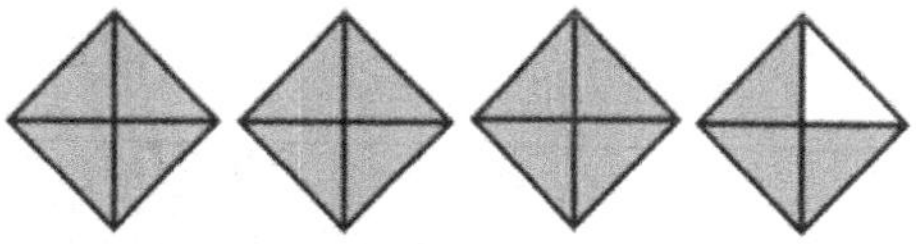 =

11) 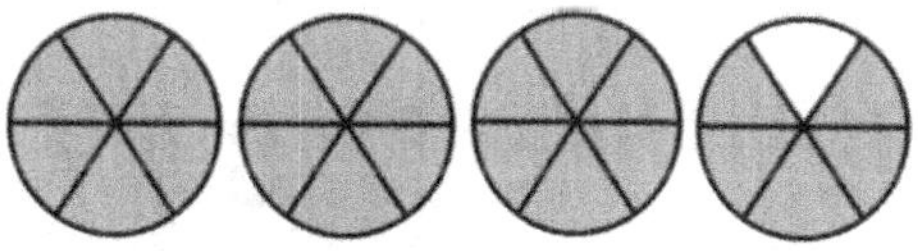 =

12) 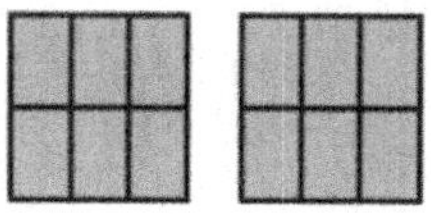 =

Name: Date: Score: Unstoppable Owl

41) Mixed Fractions to Improper Fractions

Illustrations are not needed to calculate improper fractions. Mixed fractions can be used to create improper fractions too. The denominator of the mixed fraction will be the denominator of the improper fraction. So, only the numerator needs to be calculated.

To calculate the numerator, multiply the denominator by the whole number, and add the result with any remaining parts of a whole.

Example:	Part 1: Denominator	Part 2: Calculate the numerator.
$5\frac{2}{3}$ =	$5\frac{2}{3} = \frac{\quad}{\mathbf{3}}$ *(The denominator for the irrational fraction will be the same.)*	$5\frac{2}{3} = \frac{\mathbf{17}}{\mathbf{3}}$ **Step 1:** 5 × 3 = **15** *(Multiply the denominator and whole number.)* **Step 2:** 15 + 2 = **17** *(Add the parts of the whole.)*

Directions: *Write the improper fraction for each mixed fraction.*

1) $4\frac{3}{5}$ =

2) $3\frac{1}{4}$ =

3) $8\frac{2}{7}$ =

4) $10\frac{1}{2}$ =

5) $7\frac{2}{5}$ =

6) $9\frac{1}{3}$ =

7) $11\frac{2}{5}$ =

8) $6\frac{3}{8}$ =

9) $12\frac{3}{4}$ =

10) $7\frac{2}{5}$ =

11) $4\frac{3}{8}$ =

12) $6\frac{1}{6}$ =

13) $10\frac{3}{4}$ =

14) $2\frac{10}{11}$ =

15) $8\frac{2}{3}$ =

16) $1\frac{4}{9}$ =

17) $5\frac{3}{4}$ =

18) $1\frac{4}{5}$ =

19) $8\frac{7}{10}$ =

20) $20\frac{1}{3}$ =

21) $9\frac{3}{4}$ =

22) $4\frac{2}{7}$ =

23) $7\frac{1}{2}$ =

24) $2\frac{21}{25}$ =

Name: Date: Score: Unstoppable Owl

42) Improper Fractions to Mixed Fractions

Improper fractions can also be converted to a mixed fraction by dividing the improper fraction's numerator by the denominator. This will calculate the whole number for the mixed fraction and any remainders will be the numerator for the mixed fraction. The denominator for the mixed fraction will be the same as the denominator for the improper fraction.

If there are no remainders, then the answer is not a mixed fraction. It's a whole number.

Examples:	Step 1: Divide	Answer:
$\frac{9}{2}$ =	9 ÷ 2 = **4 R1**	$\frac{9}{2} = 4\frac{1}{2}$
$\frac{21}{7}$ =	21 ÷ 7 = **3**	$\frac{21}{7} = 3$
$\frac{47}{4}$ =	47 ÷ 4 = **11 R3**	$\frac{47}{4} = 11\frac{3}{4}$

Additional resources are available if you are not comfortable with division or remainders.

Directions: *Write the mixed fraction or whole number for each improper fraction.*

1) $\frac{5}{2}$ =

2) $\frac{9}{3}$ =

3) $\frac{33}{5}$ =

4) $\frac{17}{4}$ =

5) $\frac{45}{8}$ =

6) $\frac{18}{2}$ =

7) $\frac{7}{5}$ =

8) $\frac{57}{10}$ =

9) $\frac{13}{3}$ =

10) $\frac{23}{2}$ =

11) $\frac{27}{4}$ =

12) $\frac{65}{9}$ =

13) $\frac{39}{5}$ =

14) $\frac{23}{2}$ =

15) $\frac{42}{6}$ =

16) $\frac{35}{8}$ =

17) $\frac{13}{7}$ =

18) $\frac{34}{9}$ =

19) $\frac{19}{5}$ =

20) $\frac{28}{3}$ =

21) $\frac{75}{7}$ =

22) $\frac{57}{5}$ =

23) $\frac{39}{10}$ =

24) $\frac{24}{3}$ =

Name: Date: Score: Unstoppable Owl

43) Mixed & Improper Fractions Review

Directions: *Write the improper fraction for each mixed fraction.*

1) $6\frac{1}{7}$ =

2) $1\frac{7}{12}$ =

3) $11\frac{2}{6}$ =

4) $9\frac{4}{5}$ =

5) $4\frac{3}{8}$ =

6) $15\frac{1}{2}$ =

7) $10\frac{4}{9}$ =

8) $3\frac{1}{6}$ =

9) $7\frac{3}{4}$ =

10) $5\frac{6}{7}$ =

11) $11\frac{1}{4}$ =

12) $8\frac{5}{6}$ =

13) $6\frac{1}{3}$ =

14) $25\frac{2}{3}$ =

15) $1\frac{7}{40}$ =

16) $10\frac{1}{8}$ =

17) $9\frac{3}{9}$ =

18) $2\frac{11}{25}$ =

Directions: *Write the mixed fraction or whole number for each improper fraction.*

19) $\frac{14}{3}$ =

20) $\frac{23}{8}$ =

21) $\frac{36}{9}$ =

22) $\frac{29}{5}$ =

23) $\frac{19}{2}$ =

24) $\frac{54}{6}$ =

25) $\frac{36}{7}$ =

26) $\frac{44}{11}$ =

27) $\frac{56}{9}$ =

28) $\frac{28}{5}$ =

29) $\frac{67}{10}$ =

30) $\frac{59}{5}$ =

31) $\frac{37}{12}$ =

32) $\frac{41}{2}$ =

33) $\frac{13}{8}$ =

34) $\frac{63}{25}$ =

35) $\frac{87}{10}$ =

36) $\frac{100}{20}$ =

Name: Date: Score: Unstoppable Owl

44) Comparing Mixed & Improper Fractions (Same Denominators)

When comparing mixed and improper fractions, it often helps to convert one of the fractions so that you have two mixed fractions to compare or two improper fractions to compare. Having the fractions in the same format makes it simple to compare them.

Directions: *Draw a <, >, or = symbol to compare the fractions.*

1) $\frac{11}{2}$ $8\frac{1}{2}$

2) $1\frac{3}{4}$ $\frac{6}{2}$

3) $3\frac{4}{5}$ $\frac{19}{5}$

4) $\frac{21}{3}$ $5\frac{2}{3}$

5) $7\frac{3}{8}$ $\frac{49}{8}$

6) $\frac{38}{5}$ $7\frac{3}{5}$

7) $\frac{77}{10}$ $8\frac{1}{10}$

8) $4\frac{1}{3}$ $\frac{14}{3}$

9) $\frac{20}{6}$ $2\frac{5}{6}$

10) $9\frac{7}{4}$ $\frac{35}{4}$

11) $2\frac{9}{10}$ $\frac{28}{10}$

12) $8\frac{1}{5}$ $\frac{42}{5}$

13) $\frac{7}{3}$ $2\frac{2}{3}$

14) $3\frac{1}{4}$ $\frac{12}{4}$

15) $\frac{51}{11}$ $5\frac{1}{11}$

16) $\frac{60}{7}$ $6\frac{2}{7}$

17) $1\frac{11}{15}$ $\frac{27}{15}$

18) $\frac{11}{9}$ $1\frac{1}{9}$

19) $5\frac{2}{4}$ $\frac{25}{4}$

20) $\frac{45}{8}$ $6\frac{1}{8}$

21) $\frac{26}{7}$ $3\frac{4}{7}$

22) $2\frac{1}{2}$ $\frac{6}{2}$

23) $\frac{30}{12}$ $3\frac{5}{12}$

24) $\frac{48}{5}$ $8\frac{4}{5}$

25) $10\frac{1}{3}$ $\frac{35}{3}$

26) $\frac{70}{6}$ $11\frac{5}{6}$

27) $7\frac{2}{9}$ $\frac{59}{9}$

28) $1\frac{11}{12}$ $\frac{22}{12}$

29) $\frac{30}{2}$ $14\frac{1}{2}$

30) $4\frac{4}{7}$ $\frac{31}{7}$

Name: Date: Score: Unstoppable Owl

45) Reducing Fractions

Page 10 introduced **equivalent fractions**, which are fractions that have an equal value. **Reducing fractions** (often called **simplifying fractions**) uses division to find an equivalent fraction that is stated in a simpler form. For instance, a person could say four-eighths of a pizza is left, or they could reduce this fraction and say half the pizza is left. Four-eighths and one-half represent the same amount, but it's often easier to state the fraction in its reduced form.

Dividing a fraction's numerator and denominator by their **greatest common factor** is the surest way to reduce the fraction to its simplest form. But calculating the greatest common factor can be complex and is not covered in this book. A simple way to learn to reduce fractions is to keep dividing the numerator and denominator with the same numbers until you cannot find a number that can divide both anymore.

Reduce each fraction to its simplest from.

Examples:	**Step 1:** Divide	**Step 2+:** Can you divide again?	**Answer:**
$\frac{4}{8}$	$\frac{4 \div 4}{8 \div 4} = \frac{1}{2}$	*No, the division is done.*	$\frac{1}{2}$
$\frac{18}{24}$	$\frac{18 \div 2}{24 \div 2} = \frac{9}{12}$	*Yes.* $\frac{9 \div 3}{12 \div 3} = \frac{3}{4}$	$\frac{3}{4}$

Directions: *Reduce each fraction to its simplest from.*

1) $\frac{15}{20}$

2) $\frac{4}{14}$

3) $\frac{6}{33}$

4) $\frac{16}{20}$

5) $\frac{7}{14}$

6) $\frac{8}{10}$

7) $\frac{18}{30}$

8) $\frac{20}{100}$

9) $\frac{20}{50}$

10) $\frac{15}{25}$

11) $\frac{32}{48}$

12) $\frac{6}{27}$

13) $\frac{8}{30}$

14) $\frac{12}{36}$

15) $\frac{45}{50}$

16) $\frac{26}{44}$

17) $\frac{16}{32}$

18) $\frac{30}{45}$

19) $\frac{70}{80}$

20) $\frac{28}{50}$

21) $\frac{60}{80}$

Name: Date: Score: Unstoppable Owl

46) Reducing Fractions

On the previous page, each of the fractions in the practice problems could be reduced. On this page, some will not be able to be reduced. Some of the fractions will already be in their simplest form.

How can you tell if a fraction is already in its simplest form? The numerator and denominator cannot be evenly divided by the same number.

Fractions that are in their simplest form.

$\frac{13}{20}$ $\frac{7}{9}$ $\frac{15}{16}$ $\frac{2}{7}$

Directions: *Reduce each fraction to its simplest from, or write* **SF** *if it's already in its* ***simplest form***.

1) $\frac{12}{30}$

2) $\frac{11}{18}$

3) $\frac{6}{13}$

4) $\frac{15}{21}$

5) $\frac{8}{22}$

6) $\frac{9}{16}$

7) $\frac{23}{30}$

8) $\frac{14}{26}$

9) $\frac{5}{28}$

10) $\frac{10}{28}$

11) $\frac{18}{25}$

12) $\frac{16}{18}$

13) $\frac{6}{48}$

14) $\frac{75}{100}$

15) $\frac{7}{12}$

16) $\frac{4}{28}$

17) $\frac{10}{12}$

18) $\frac{17}{24}$

19) $\frac{8}{11}$

20) $\frac{6}{15}$

21) $\frac{34}{40}$

22) $\frac{12}{17}$

23) $\frac{8}{15}$

24) $\frac{18}{32}$

25) $\frac{7}{21}$

26) $\frac{8}{42}$

27) $\frac{6}{30}$

28) $\frac{14}{25}$

29) $\frac{13}{26}$

30) $\frac{15}{19}$

Name: Date: Score: Unstoppable Owl

47) Reducing Mixed Fractions

Reducing mixed fractions is not much different than reducing other fractions. In fact, you don't need to even focus on the whole number. Focus on the fraction portion. Find a number that can divide both the numerator and the denominator. Keep doing this until the fraction is in simplest form. As you can see in the examples below, the whole number remains the same.

Reduce each mixed fraction to its simplest from.

Examples:	**Step 1:** Divide	**Step 2+:** Can you divide again?	**Answer:**
$8\frac{12}{18}$	$\frac{12 \div 6}{18 \div 6} = \frac{2}{3}$	*No, the division is done.*	$8\frac{2}{3}$
$2\frac{10}{60}$	$\frac{10 \div 2}{60 \div 2} = \frac{5}{30}$	*Yes.* $\frac{5 \div 5}{30 \div 5} = \frac{1}{6}$	$2\frac{1}{6}$

Directions: *Reduce each mixed fraction to its simplest form.*

1) $7\frac{15}{60}$

2) $2\frac{8}{44}$

3) $9\frac{24}{50}$

4) $12\frac{14}{35}$

5) $5\frac{46}{60}$

6) $8\frac{88}{100}$

7) $10\frac{24}{80}$

8) $1\frac{3}{9}$

9) $14\frac{18}{28}$

10) $7\frac{36}{50}$

11) $5\frac{30}{45}$

12) $8\frac{40}{75}$

13) $17\frac{12}{20}$

14) $4\frac{25}{75}$

15) $6\frac{40}{90}$

16) $3\frac{6}{27}$

17) $2\frac{16}{64}$

18) $9\frac{27}{72}$

19) $4\frac{8}{50}$

20) $30\frac{9}{24}$

21) $1\frac{4}{26}$

22) $7\frac{15}{24}$

23) $5\frac{4}{32}$

24) $8\frac{56}{88}$

Name: Date: Score: Unstoppable Owl

48) Reducing Mixed Fractions

This page has improper fractions. The goal is to convert the improper fractions to mixed fractions, then simplify the mixed fractions to their simplest form.

Examples:	**Step 1:** Convert to a mixed fraction	**Step 2:** Simplify the fraction.
$\frac{18}{4}$	$\frac{18}{4} = 4\frac{2}{4}$	$4\frac{1}{2}$
$\frac{51}{9}$	$\frac{51}{9} = 5\frac{6}{9}$	$5\frac{2}{3}$

Directions: *Convert each improper fraction to a mixed fraction, then simplify.*

1) $\frac{10}{4}$

2) $\frac{26}{6}$

3) $\frac{70}{8}$

4) $\frac{84}{8}$

5) $\frac{33}{6}$

6) $\frac{40}{12}$

7) $\frac{35}{10}$

8) $\frac{62}{10}$

9) $\frac{30}{14}$

10) $\frac{21}{14}$

11) $\frac{54}{12}$

12) $\frac{75}{9}$

13) $\frac{50}{20}$

14) $\frac{95}{10}$

15) $\frac{39}{15}$

16) $\frac{84}{20}$

17) $\frac{32}{12}$

18) $\frac{102}{20}$

19) $\frac{56}{6}$

20) $\frac{90}{25}$

21) $\frac{76}{20}$

22) $\frac{66}{15}$

23) $\frac{38}{8}$

24) $\frac{35}{20}$

25) $\frac{220}{50}$

26) $\frac{146}{10}$

27) $\frac{34}{12}$

28) $\frac{22}{16}$

29) $\frac{70}{25}$

30) $\frac{44}{16}$

Name: Date: Score: Unstoppable Owl

49) Adding Decimals

Decimals can be added, much the same way multi-digit whole numbers can be. To add decimals, write the two decimals one on top of the other. Make sure that the place values are **aligned**. An easy way to do this is to make sure that the decimal point of the first number is directly above the decimal point of the second number.

Once the numbers are written, add the columns, regroup if necessary, and place the decimal in the answer. If you don't know how to add multi-digit whole numbers, don't worry. Other resources are available to practice these skills.

Decimals aligned	Decimals not aligned
23.75 + 4.48	23.75 +4.48

Examples:

1 1 23.75 + 4.48 28.23	1 0.62 + 0.73 1.35	3.4 + 51.5 54.9

Directions: *Add each set of decimals.*

1) 14.21 + 3.56	2) 8.1 + 37.5	3) 42.57 + 90.31	4) 5.71 + 6.94
5) 3.281 + 1.709	6) 0.58 + 6.29	7) 2.514 + 3.095	8) 71.3 + 84.6
9) 25.8 + 1.4	10) 8.931 + 5.037	11) 55.2 + 3.7	12) 0.2 + 0.6
13) 5.91 + 1.56	14) 2.116 + 2.528	15) 147.3 + 34.2	16) 6.85 + 42.01

Name: Date: Score: Unstoppable Owl

50) Adding Decimals

On the previous page, the problems were written vertically and aligned for you. On this page, you will need to write the numbers yourself. Place one number on top of the other. Remember to align the decimals so that the place values form columns for you to add.

Directions: *Rewrite each set of decimals vertically and then add them.*

1) **5.91 + 0.63 =**

2) **53.7 + 4.6 =**

3) **12.48 + 30.55 =**

4) **0.407 + 0.289 =**

5) **3.64 + 71.39 =**

6) **7.82 + 7.35 =**

7) **8.9 + 4.2 =**

8) **54.7 + 94.2 =**

9) **0.76 + 3.41 =**

10) **9.601 + 4.225 =**

11) **158.9 + 3.5 =**

12) **6.08 + 40.59 =**

13) **0.7 + 1.4 =**

14) **57.34 + 52.77 =**

15) **3.005 + 4.623 =**

Name: Date: Score: Unstoppable Owl

51) Adding Decimals

When decimals are being added, they may not always have an equal number of digits to the right of the decimal point. It's still important to align the decimal points before adding. But some place value columns may not have two digits to add. There will be blank spaces that you may fill with zeros. The zeros act as **placeholders** that help some people to add more accurately. You don't have to write zeros, but they may be helpful.

Example:	**Option 1:** Leave the blank spaces.	**Option 2:** Write the zeros.
4.7 + 2.165 =	$\begin{array}{r} 4.7 \\ +\ 2.165 \\ \hline 6.865 \end{array}$	$\begin{array}{r} 4.700 \\ +\ 2.165 \\ \hline 6.865 \end{array}$

As you can see, the example's answer is the same whether you choose to write the zeros or not. Try both options on the problems below and decide which option you like the best.

Directions: *Rewrite each set of decimals vertically and then add them.*

1) **5.9 + 6.48 =**

2) **2.74 + 80.5 =**

3) **0.016 + 34.1 =**

4) **6.617 + 3.57 =**

5) **75.64 + 19.315 =**

6) **0.8251 + 0.04 =**

7) **9.3 + 1.886 =**

8) **250.73 + 94.2 =**

9) **3.7 + 34.61 =**

10) **18.6 + 4.9538 =**

11) **1.98261 + 7.24 =**

12) **6.4219 + 40.5 =**

Name: Date: Score: Unstoppable Owl

52) Adding Decimals

Sometimes, fractions need to be added with a whole number. *How do you line up the decimals then?* The decimal for a whole number would be to the right of the whole number. You may write this decimal and you may even write zeros to the right of the decimal help keep track of the place values.

Example:	**Option 1:** Leave the blank spaces.	**Option 2:** Write the zeros.
6 + 2.938 =	$\begin{array}{r} 6 \\ +\ 2.938 \\ \hline 8.938 \end{array}$	$\begin{array}{r} 6.000 \\ +\ 2.938 \\ \hline 8.938 \end{array}$

Directions: *Rewrite each set of decimals vertically and then add them.*

1) **7 + 4.95 =**

2) **4.12 + 5 =**

3) **3 + 16.91 =**

4) **82 + 5.29 =**

5) **75.4 + 16 =**

6) **163 + 7.5 =**

7) **0.873 + 6 =**

8) **56 + 381.2 =**

9) **0.0508 + 7 =**

10) **185 + 4.58 =**

11) **2.4 + 60 =**

12) **31 + 2.405 =**

Name: Date: Score: Unstoppable Owl

53) Adding Decimals

Directions: *Rewrite each set of decimals vertically and then add them.*

1) **7.61 + 3.45 =**

2) **28.7 + 4.29 =**

3) **105 + 5.78 =**

4) **0.7 + 6.93 =**

5) **8 + 2.309 =**

6) **9.82 + 0.4 =**

7) **2.076 + 6.5 =**

8) **200.6 + 56.14 =**

9) **0.718 + 5.3 =**

10) **45.2 + 6.58 =**

11) **0.9 + 0.325 =**

12) **6.07 + 1.539 =**

13) **924.7 + 2.8 =**

14) **2.8 + 1.75 =**

15) **61.5 + 7.8 =**

16) **0.31 + 84.7 =**

17) **374.34 + 8.37 =**

18) **3.172 + 4.685 =**

Name: Date: Score: Unstoppable Owl

54) Subtracting Decimals

Subtracting decimals is very similar to adding decimals. It's important to align the decimal points and place values. After that, subtract each column and **regroup** when necessary.

Example 1:	**Written Vertically:**	**Example 2:**	**Written Vertically:**
7.943 – 2.582 =	8 14 7.~~9~~~~4~~3 – 2.582 5.361	**42.5 – 2.16 =**	4 10 42.~~5~~~~0~~ – 2.16 40.34

Directions: *Rewrite each set of decimals vertically and then subtract them.*

1) **4.6 – 1.7 =**

2) **25.9 – 8.2 =**

3) **8.529 – 3.48 =**

4) **7.5 – 4.12 =**

5) **4.63 – 4.24 =**

6) **126.4 – 9.7 =**

7) **50.3 – 21.85 =**

8) **8.298 – 6.527 =**

9) **1.703 – 0.34 =**

10) **53 – 5.8 =**

11) **6 – 0.52 =**

12) **9 – 2.047 =**

Name: Date: Score: Unstoppable Owl

55) Subtracting Decimals

Directions: *Rewrite each set of decimals vertically and then subtract them.*

1) **25.3 – 8.6 =**

2) **4.81 – 1.56 =**

3) **73.28 – 16.7 =**

4) **40 – 5.7 =**

5) **0.647 – 0.085 =**

6) **6.19 – 0.852 =**

7) **512.3 – 361.5 =**

8) **7.08 – 4.25 =**

9) **13 – 7.39 =**

10) **3.6 – 3.27 =**

11) **6.588 – 1.736 =**

12) **0.075 – 0.029 =**

13) **60.45 – 3.7 =**

14) **2.6 – 1.305 =**

15) **237.6 – 9.47 =**

16) **0.685 – 0.592 =**

17) **16 – 7.413 =**

18) **5.675 – 3.8 =**

Name: Date: Score: Unstoppable Owl

56) Subtracting Decimals

Directions: *Rewrite each set of decimals vertically and then subtract them.*

1) **7.2 – 2.5 =**

2) **8.16 – 1.79 =**

3) **9.4 – 3.182 =**

4) **65.3 – 28.5 =**

5) **20.3 – 8.15 =**

6) **57 – 4.36 =**

7) **5 – 1.32 =**

8) **6.2 – 4.405 =**

9) **1.542 – 0.8 =**

10) **36.79 – 9.51 =**

11) **22.6 – 5.9 =**

12) **8.27 – 7.583 =**

13) **5.2 – 1.5 =**

14) **40.3 – 0.65 =**

15) **0.064 – 0.009 =**

16) **156.2 – 7.4 =**

17) **26 – 15.37 =**

18) **1 – 0.623 =**

Name: Date: Score: Unstoppable Owl

57) Adding Fractions (With Like Denominators)

Adding fractions is easy when the fractions have the same denominator. Just add the numerators together. The denominator for the answer will remain the same.

Examples:

$\frac{3}{7} + \frac{2}{7} = \frac{5}{7}$ | $\frac{5}{6} + \frac{3}{6} = \frac{8}{6}$ | $\frac{10}{11} + \frac{7}{11} = \frac{17}{11}$

Directions: *Add each set of fractions. If more than a whole, write answers as improper fractions.*

1) $\frac{4}{9} + \frac{3}{9} =$

2) $\frac{1}{3} + \frac{1}{3} =$

3) $\frac{5}{6} + \frac{3}{6} =$

4) $\frac{7}{2} + \frac{4}{2} =$

5) $\frac{4}{5} + \frac{4}{5} =$

6) $\frac{1}{8} + \frac{2}{8} =$

7) $\frac{4}{15} + \frac{6}{15} =$

8) $\frac{3}{4} + \frac{9}{4} =$

9) $\frac{3}{10} + \frac{1}{10} =$

10) $\frac{8}{7} + \frac{6}{7} =$

11) $\frac{9}{2} + \frac{5}{2} =$

12) $\frac{2}{12} + \frac{7}{12} =$

13) $\frac{6}{5} + \frac{10}{5} =$

14) $\frac{9}{7} + \frac{9}{7} =$

15) $\frac{4}{9} + \frac{1}{9} =$

16) $\frac{1}{20} + \frac{14}{20} =$

17) $\frac{10}{6} + \frac{25}{6} =$

18) $\frac{6}{14} + \frac{5}{14} =$

19) $\frac{5}{3} + \frac{2}{3} =$

20) $\frac{9}{25} + \frac{12}{25} =$

21) $\frac{2}{30} + \frac{2}{30} =$

22) $\frac{3}{8} + \frac{5}{8} =$

23) $\frac{7}{4} + \frac{20}{4} =$

24) $\frac{9}{10} + \frac{4}{10} =$

25) $\frac{2}{5} + \frac{7}{5} =$

26) $\frac{8}{100} + \frac{8}{100} =$

27) $\frac{1}{2} + \frac{2}{2} =$

28) $\frac{4}{6} + \frac{13}{6} =$

29) $\frac{40}{45} + \frac{30}{45} =$

30) $\frac{5}{18} + \frac{7}{18} =$

Name: Date: Score: Unstoppable Owl

58) Subtracting Fractions (With Like Denominators)

Subtracting fractions with like denominators is like adding fractions with like denominators. Subtract the numerators. The denominator for the answer will remain the same.

Examples:

$\frac{7}{8} - \frac{2}{8} = \frac{5}{8}$ | $\frac{5}{9} - \frac{1}{9} = \frac{4}{9}$ | $\frac{13}{15} - \frac{2}{15} = \frac{11}{15}$

Directions: *Subtract the fractions. If more than a whole, write answers as improper fractions.*

1) $\frac{5}{6} - \frac{3}{6} =$

2) $\frac{2}{4} - \frac{1}{4} =$

3) $\frac{14}{9} - \frac{2}{9} =$

4) $\frac{7}{3} - \frac{5}{3} =$

5) $\frac{15}{12} - \frac{8}{12} =$

6) $\frac{20}{7} - \frac{6}{7} =$

7) $\frac{9}{2} - \frac{2}{2} =$

8) $\frac{4}{5} - \frac{2}{5} =$

9) $\frac{11}{8} - \frac{7}{8} =$

10) $\frac{3}{10} - \frac{1}{10} =$

11) $\frac{17}{7} - \frac{1}{7} =$

12) $\frac{11}{3} - \frac{4}{3} =$

13) $\frac{11}{20} - \frac{10}{20} =$

14) $\frac{7}{16} - \frac{2}{16} =$

15) $\frac{68}{5} - \frac{13}{5} =$

16) $\frac{19}{11} - \frac{12}{11} =$

17) $\frac{25}{9} - \frac{15}{9} =$

18) $\frac{9}{4} - \frac{7}{4} =$

19) $\frac{2}{2} - \frac{1}{2} =$

20) $\frac{27}{35} - \frac{10}{35} =$

21) $\frac{30}{5} - \frac{7}{5} =$

22) $\frac{50}{70} - \frac{20}{70} =$

23) $\frac{15}{8} - \frac{12}{8} =$

24) $\frac{30}{12} - \frac{25}{12} =$

25) $\frac{5}{6} - \frac{1}{6} =$

26) $\frac{42}{50} - \frac{10}{50} =$

27) $\frac{80}{7} - \frac{1}{7} =$

28) $\frac{58}{100} - \frac{5}{100} =$

29) $\frac{90}{28} - \frac{40}{28} =$

30) $\frac{7}{60} - \frac{6}{60} =$

Name: Date: Score: Unstoppable Owl

59) Adding Mixed Fractions (With Like Denominators)

One way to add mixed fractions that have like denominators is to convert mixed fractions to improper fractions before adding. The answer can then be converted to a mixed fraction. (Whole numbers can be converted to an improper fraction too.)

Examples:	**Step 1:** (Convert)	**Step 2:** (Add)	**Step 3:** (Convert)
$4\frac{1}{5} + 3\frac{2}{5} =$	$\frac{21}{5} + \frac{17}{5} =$	$\frac{21}{5} + \frac{17}{5} = \frac{38}{5}$	$7\frac{3}{5}$
$3\frac{2}{3} + 2\frac{2}{3} =$	$\frac{11}{3} + \frac{8}{3} =$	$\frac{11}{3} + \frac{8}{3} = \frac{19}{3}$	$6\frac{1}{3}$

Directions: *Add the mixed fractions and state your answer as a mixed fraction.*

1) $1\frac{3}{6} + 3\frac{2}{6} =$

2) $4\frac{1}{4} + 4\frac{2}{4} =$

3) $5\frac{6}{7} + 2\frac{5}{7} =$

4) $3\frac{8}{10} + 3\frac{7}{10} =$

5) $4\frac{2}{5} + 11\frac{2}{5} =$

6) $1\frac{8}{9} + 1\frac{3}{9} =$

7) $7\frac{5}{8} + 6 =$

8) $2\frac{5}{6} + 2\frac{4}{6} =$

9) $5\frac{3}{11} + \frac{2}{11} =$

10) $4\frac{7}{20} + 5\frac{11}{20} =$

11) $8\frac{1}{2} + 3\frac{1}{2} =$

12) $\frac{3}{5} + 7\frac{3}{5} =$

13) $7\frac{5}{7} + 9\frac{4}{7} =$

14) $1\frac{1}{3} + 4\frac{2}{3} =$

15) $5\frac{6}{8} + \frac{1}{8} =$

16) $3\frac{7}{10} + 8\frac{9}{10} =$

17) $4 + 2\frac{3}{4} =$

18) $6\frac{4}{7} + 1\frac{2}{7} =$

19) $\frac{1}{3} + 1\frac{1}{3} =$

20) $2\frac{10}{16} + 1\frac{10}{16} =$

21) $13 + 3\frac{7}{9} =$

22) $6\frac{5}{14} + 2\frac{8}{14} =$

23) $4\frac{2}{4} + \frac{3}{4} =$

24) $1\frac{4}{6} + 7\frac{2}{6} =$

25) $3\frac{13}{20} + 2\frac{14}{20} =$

26) $4\frac{7}{10} + 27 =$

27) $3\frac{3}{4} + 5\frac{3}{4} =$

Name: Date: Score: Unstoppable Owl

60) Adding Mixed Fractions (With Like Denominators)

Another way to add mixed fractions is to add the fraction portions of the mixed fractions, then add the whole numbers. Sometimes, the fraction portions will equal more than one whole. When this happens, the one whole needs to be added with the other whole numbers.

Examples:	(Add Fractions)	(Add Whole Numbers)	(Answer)
$4\frac{1}{5} + 3\frac{2}{5} =$	$\frac{1}{5} + \frac{2}{5} = \frac{3}{5}$	$4 + 3 = 7$	$7\frac{3}{5}$
$3\frac{2}{3} + 2\frac{2}{3} =$	$\frac{2}{3} + \frac{2}{3} = 1\frac{1}{3}$	$3 + 2 + 1 = 6$	$6\frac{1}{3}$

Directions: *Add the mixed fractions and state your answer as a mixed fraction.*

1) $1\frac{3}{7} + 4\frac{2}{7} =$

2) $2\frac{3}{4} + 1\frac{1}{4} =$

3) $4\frac{4}{8} + 3\frac{5}{8} =$

4) $6\frac{2}{3} + 2\frac{1}{3} =$

5) $3\frac{4}{6} + 41\frac{3}{6} =$

6) $2\frac{7}{10} + 9\frac{9}{10} =$

7) $3\frac{1}{2} + 5\frac{1}{2} =$

8) $2\frac{6}{7} + \frac{6}{7} =$

9) $4\frac{13}{15} + 8\frac{8}{15} =$

10) $4\frac{2}{5} + 5 =$

11) $7\frac{5}{8} + 3\frac{2}{8} =$

12) $4\frac{9}{12} + \frac{8}{12} =$

13) $2\frac{4}{5} + 5\frac{4}{5} =$

14) $10 + 1\frac{2}{9} =$

15) $4\frac{1}{5} + 2\frac{2}{5} =$

16) $8\frac{5}{7} + 3 =$

17) $1\frac{17}{20} + 1\frac{15}{20} =$

18) $6\frac{5}{8} + \frac{7}{8} =$

19) $3\frac{3}{4} + 3\frac{3}{4} =$

20) $2\frac{3}{9} + 1\frac{4}{9} =$

21) $3\frac{1}{7} + 5\frac{3}{7} =$

22) $2\frac{23}{25} + 25\frac{7}{25} =$

23) $1\frac{5}{6} + 6\frac{5}{6} =$

24) $7\frac{12}{15} + 5\frac{14}{15} =$

25) $9\frac{7}{9} + 9\frac{1}{9} =$

26) $5\frac{8}{11} + 2\frac{10}{11} =$

27) $7\frac{9}{14} + 1\frac{5}{14} =$

Name: Date: Score: Unstopable Owl

61) Adding Mixed Fractions (With Like Denominators)

You've learned two ways to add mixed fractions. The method that adds the fractions then adds the whole numbers is often easier—especially when large whole numbers are involved. For instance, in the first problem below, changing sixty-four and one-seventh to an improper fraction may be difficult. But adding the fraction portion and the whole numbers separately is fairly easy. Use whichever method you prefer to add the fractions below.

Directions: *Add the mixed fractions and state your answer as a mixed fraction.*

1) $3\frac{5}{7} + 64\frac{1}{7} =$

2) $9\frac{3}{6} + 8\frac{4}{6} =$

3) $38\frac{2}{3} + 5\frac{2}{3} =$

4) $6\frac{4}{5} + 7\frac{3}{5} =$

5) $2\frac{4}{10} + 92\frac{9}{10} =$

6) $8\frac{2}{4} + 8\frac{3}{4} =$

7) $9\frac{8}{12} + 57\frac{11}{12} =$

8) $\frac{30}{75} + 9\frac{50}{75} =$

9) $6\frac{1}{5} + 1\frac{4}{5} =$

10) $8\frac{7}{8} + 5\frac{7}{8} =$

11) $9\frac{11}{14} + 9\frac{12}{14} =$

12) $6\frac{12}{25} + 43\frac{14}{25} =$

13) $\frac{2}{9} + 55\frac{5}{9} =$

14) $8\frac{9}{11} + 6\frac{7}{11} =$

15) $1\frac{5}{7} + 1\frac{5}{7} =$

16) $27\frac{8}{9} + 3\frac{5}{9} =$

17) $4\frac{2}{5} + 3\frac{2}{5} =$

18) $2\frac{5}{8} + 7\frac{3}{8} =$

19) $8\frac{15}{17} + 36\frac{6}{17} =$

20) $4\frac{7}{15} + 7\frac{10}{15} =$

21) $80\frac{2}{6} + \frac{5}{6} =$

22) $5\frac{7}{30} + 4\frac{19}{30} =$

23) $3\frac{1}{5} + 67\frac{2}{5} =$

24) $1\frac{24}{40} + 5\frac{25}{40} =$

25) $\frac{2}{9} + 29\frac{7}{9} =$

26) $8\frac{12}{13} + 3\frac{8}{13} =$

27) $2\frac{1}{7} + 6\frac{6}{7} =$

28) $44 + 3\frac{2}{3} =$

29) $3\frac{4}{6} + 2\frac{2}{6} =$

30) $7\frac{7}{18} + 1\frac{17}{18} =$

Name: Date: Score: Unstoppable Owl

62) Subtracting Mixed Fractions (With Like Denominators)

There are two common ways to subtract mixed fractions that have the same denominator. The mixed fractions can be converted to improper fractions before subtracting, or the fractions and whole numbers can be subtracted separately. (Similar to adding, the second option is often easier. But, you may choose which method you prefer.)

With subtracting there may be one additional step. If the fraction portion of the number that is being subtracted from is smaller than the fraction portion of the number being subtracted, then you need to regroup. You can regroup one from the whole number and add it to the fraction portion to create an improper fraction.

Examples:	**Regroup** (If needed.)	**− Fractions**	**− Whole Numbers**	(Answer)
$6\frac{4}{5} - 2\frac{3}{5} =$	**Not Needed.**	$\frac{4}{5} - \frac{3}{5} = \frac{1}{5}$	$6 - 2 = 4$	$4\frac{1}{5}$
$8\frac{2}{8} - 5\frac{6}{8} =$	$7\frac{10}{8} - 5\frac{6}{8} =$	$\frac{10}{8} - \frac{6}{8} = \frac{4}{8}$	$7 - 5 = 2$	$2\frac{4}{8}$

Directions: *Subtract the mixed fractions and state your answer as a mixed fraction.*

1) $5\frac{4}{5} - 3\frac{2}{5} =$

2) $4\frac{5}{7} - 1\frac{4}{7} =$

3) $8\frac{1}{4} - 6\frac{3}{4} =$

4) $7\frac{3}{5} - 2\frac{4}{5} =$

5) $13\frac{5}{9} - 1\frac{8}{9} =$

6) $2\frac{1}{3} - 1\frac{2}{3} =$

7) $9\frac{7}{10} - 5\frac{1}{10} =$

8) $8\frac{3}{12} - 2\frac{7}{12} =$

9) $7\frac{1}{4} - 6\frac{2}{4} =$

10) $3\frac{1}{6} - \frac{5}{6} =$

11) $6\frac{1}{2} - 3\frac{1}{2} =$

12) $9\frac{5}{7} - 2\frac{4}{7} =$

13) $5\frac{2}{8} - 4\frac{7}{8} =$

14) $8\frac{9}{11} - 1\frac{4}{11} =$

15) $3\frac{6}{7} - 1\frac{4}{7} =$

16) $8\frac{8}{11} - 4\frac{1}{11} =$

17) $6\frac{3}{15} - 5\frac{14}{15} =$

18) $23\frac{1}{6} - 8\frac{4}{6} =$

19) $6\frac{3}{20} - \frac{9}{20} =$

20) $7\frac{5}{7} - 3\frac{5}{7} =$

21) $5\frac{23}{30} - 2\frac{4}{30} =$

Name: Date: Score: Unstoppable Owl

63) Adding Fractions (With Unlike Denominators)

Adding fractions is easy when the fractions have the same denominator. *But what if they don't have the same denominator?* When fractions have unlike denominators, most people write equivalent fraction that allows them to have **common denominators** before adding.

Different people may choose to create different equivalent fractions that use different common denominators before adding. For this reason, it is best to simplify your answer after adding, so that any fractions will be stated in their simplest form.

Examples:	Rewritten	Added	Simplified Answer
$\frac{1}{2} + \frac{1}{6} =$	$\frac{3}{6} + \frac{1}{6} =$	$\frac{3}{6} + \frac{1}{6} = \frac{4}{6}$	$\frac{4}{6} = \frac{2}{3}$
$\frac{3}{8} + \frac{2}{4} =$	$\frac{3}{8} + \frac{4}{8} =$	$\frac{3}{8} + \frac{4}{8} = \frac{7}{8}$	$\frac{7}{8}$ is already simplified.

Directions: *Add each set of fractions. If more than a whole, write answers as improper fractions.*

1) $\frac{1}{2} + \frac{1}{8} =$

2) $\frac{2}{12} + \frac{2}{3} =$

3) $\frac{2}{3} + \frac{1}{6} =$

4) $\frac{1}{5} + \frac{3}{10} =$

5) $\frac{1}{20} + \frac{3}{4} =$

6) $\frac{2}{5} + \frac{4}{15} =$

7) $\frac{7}{40} + \frac{1}{5} =$

8) $\frac{3}{14} + \frac{1}{2} =$

9) $\frac{1}{30} + \frac{3}{15} =$

10) $\frac{5}{12} + \frac{1}{2} =$

11) $\frac{1}{6} + \frac{13}{24} =$

12) $\frac{1}{16} + \frac{3}{4} =$

13) $\frac{1}{3} + \frac{2}{9} =$

14) $\frac{6}{25} + \frac{2}{5} =$

15) $\frac{11}{40} + \frac{7}{20} =$

16) $\frac{5}{18} + \frac{3}{6} =$

17) $\frac{1}{24} + \frac{5}{8} =$

18) $\frac{1}{20} + \frac{7}{10} =$

19) $\frac{1}{4} + \frac{13}{28} =$

20) $\frac{1}{3} + \frac{5}{12} =$

21) $\frac{6}{20} + \frac{3}{5} =$

22) $\frac{2}{25} + \frac{32}{100} =$

23) $\frac{3}{16} + \frac{1}{2} =$

24) $\frac{2}{9} + \frac{11}{18} =$

Name: Date: Score: Unstoppable Owl

64) Adding Fractions (With Unlike Denominators)

On the previous page, only one of the fractions needed to be rewritten as an equivalent fraction to make the denominators the same. But sometimes, both fractions need to be rewritten to create a common denominator.

In the first example, the fractions were rewritten using 12 as a common denominator. Other people may have chosen to use 24 as the common denominator, since 4 x 6 = 24. It doesn't matter which common denominator you use—if the answer is simplified after adding, the answer will be the same. (See for yourself. 24 was used in the second example.)

Examples:	Rewritten	Added	Simplified Answer
$\frac{1}{4} + \frac{1}{6} =$	$\frac{3}{12} + \frac{2}{12} =$	$\frac{3}{12} + \frac{2}{12} = \frac{5}{12}$	$\frac{5}{12}$ is already simplified.
$\frac{1}{4} + \frac{1}{6} =$	$\frac{6}{24} + \frac{4}{24} =$	$\frac{6}{24} + \frac{4}{24} = \frac{10}{24}$	$\frac{10}{24} = \frac{5}{12}$
$\frac{2}{10} + \frac{3}{4} =$	$\frac{8}{40} + \frac{30}{40} =$	$\frac{8}{40} + \frac{30}{40} = \frac{38}{40}$	$\frac{38}{40} = \frac{19}{20}$

Directions: *Add each set of fractions. If more than a whole, write answers as improper fractions.*

1) $\frac{2}{3} + \frac{1}{4} =$

2) $\frac{2}{5} + \frac{1}{3} =$

3) $\frac{2}{4} + \frac{2}{6} =$

4) $\frac{1}{8} + \frac{3}{6} =$

5) $\frac{3}{10} + \frac{1}{3} =$

6) $\frac{2}{5} + \frac{4}{6} =$

7) $\frac{3}{7} + \frac{1}{2} =$

8) $\frac{2}{6} + \frac{2}{8} =$

9) $\frac{3}{4} + \frac{1}{10} =$

10) $\frac{5}{8} + \frac{1}{3} =$

11) $\frac{1}{4} + \frac{3}{7} =$

12) $\frac{1}{3} + \frac{1}{2} =$

13) $\frac{2}{5} + \frac{2}{4} =$

14) $\frac{1}{6} + \frac{5}{9} =$

15) $\frac{1}{4} + \frac{3}{5} =$

16) $\frac{1}{3} + \frac{1}{4} =$

17) $\frac{4}{9} + \frac{2}{6} =$

18) $\frac{2}{6} + \frac{2}{5} =$

19) $\frac{1}{2} + \frac{2}{9} =$

20) $\frac{1}{3} + \frac{2}{5} =$

21) $\frac{5}{12} + \frac{3}{8} =$

Name: Date: Score: Unstoppable Owl

65) Adding Improper Fractions (With Unlike Denominators)

You can add improper fractions with unlike denominators the same way you have been adding other fractions with unlike denominators. Rewrite the fractions as equivalent fractions so that the denominators are the same. Once the denominators are the same, add the numerators. Then, simplify if necessary.

Examples:	**Rewritten**	**Added**	**Simplified Answer**
$\frac{5}{3} + \frac{7}{6} =$	$\frac{10}{6} + \frac{7}{6} =$	$\frac{10}{6} + \frac{7}{6} = \frac{17}{6}$	$\frac{17}{6}$ is already simplified.
$\frac{3}{12} + \frac{9}{4} =$	$\frac{3}{12} + \frac{27}{12} =$	$\frac{3}{12} + \frac{27}{12} = \frac{30}{12}$	$\frac{30}{12} = \frac{5}{2}$

Directions: *Add each set of fractions and simplify your answer.*
If more than a whole, write answers as improper fractions in simplified form.

1) $\frac{7}{4} + \frac{11}{8} =$

2) $\frac{3}{2} + \frac{8}{3} =$

3) $\frac{5}{3} + \frac{14}{9} =$

4) $\frac{7}{5} + \frac{13}{10} =$

5) $\frac{4}{3} + \frac{9}{4} =$

6) $\frac{9}{8} + \frac{10}{6} =$

7) $\frac{7}{20} + \frac{7}{4} =$

8) $\frac{20}{7} + \frac{1}{2} =$

9) $\frac{3}{2} + \frac{15}{9} =$

10) $\frac{10}{8} + \frac{6}{5} =$

11) $\frac{5}{12} + \frac{11}{8} =$

12) $\frac{13}{6} + \frac{4}{3} =$

13) $\frac{7}{6} + \frac{5}{9} =$

14) $\frac{8}{3} + \frac{20}{12} =$

15) $\frac{10}{8} + \frac{9}{2} =$

16) $\frac{9}{4} + \frac{7}{6} =$

17) $\frac{11}{2} + \frac{7}{10} =$

18) $\frac{8}{5} + \frac{9}{7} =$

19) $\frac{5}{2} + \frac{11}{5} =$

20) $\frac{12}{11} + \frac{7}{4} =$

21) $\frac{10}{3} + \frac{5}{2} =$

22) $\frac{9}{5} + \frac{7}{6} =$

23) $\frac{6}{5} + \frac{17}{15} =$

24) $\frac{9}{4} + \frac{10}{9} =$

25) $\frac{9}{8} + \frac{9}{2} =$

26) $\frac{5}{3} + \frac{20}{8} =$

27) $\frac{25}{15} + \frac{10}{6} =$

Name: Date: Score: Unstoppable Owl

66) Adding Mixed Fractions (With Unlike Denominators)

Adding mixed fractions was introduced on page 59. The last few pages focused on adding fractions with unlike denominators. Now, it's time to combine these skills and add mixed fractions with unlike denominators. The examples below show the method where the fractions and whole numbers are added separately.

Examples:	(Rewrite Fractions then +)	(+ Whole Numbers)	(Simplified Answer)
$3\frac{1}{4} + 5\frac{3}{8} =$	$\frac{2}{8} + \frac{3}{8} = \frac{5}{8}$	$3 + 5 = 8$	$8\frac{5}{8}$ is already simplified.
$2\frac{5}{6} + 4\frac{2}{3} =$	$\frac{5}{6} + \frac{4}{6} = 1\frac{3}{6}$	$2 + 4 + 1 = 7$	$7\frac{3}{6} = 7\frac{1}{2}$

Directions: *Add the mixed fractions and state your answer as a mixed fraction in simplified form.*

1) $5\frac{5}{8} + 14\frac{1}{2} =$

2) $7\frac{2}{3} + 8\frac{3}{5} =$

3) $23\frac{2}{6} + 9\frac{1}{4} =$

4) $1\frac{8}{12} + 2\frac{1}{3} =$

5) $6\frac{2}{5} + \frac{5}{6} =$

6) $4\frac{1}{2} + 1\frac{5}{9} =$

7) $8\frac{7}{15} + 31\frac{2}{3} =$

8) $\frac{5}{9} + 5\frac{2}{6} =$

9) $3\frac{1}{2} + 3\frac{2}{5} =$

10) $7\frac{4}{7} + 1\frac{1}{4} =$

11) $4\frac{4}{5} + 10\frac{6}{8} =$

12) $\frac{3}{6} + 99\frac{1}{2} =$

13) $3\frac{3}{10} + 5\frac{2}{3} =$

14) $2\frac{2}{3} + 2\frac{5}{8} =$

15) $60\frac{8}{9} + 5\frac{1}{4} =$

16) $2\frac{2}{5} + 4\frac{2}{6} =$

17) $6\frac{5}{6} + 1\frac{3}{8} =$

18) $7\frac{2}{5} + 71\frac{6}{10} =$

19) $24\frac{3}{5} + \frac{1}{6} =$

20) $1\frac{2}{9} + 1\frac{3}{4} =$

21) $8\frac{2}{3} + 26\frac{4}{7} =$

22) $6\frac{5}{6} + 4\frac{5}{9} =$

23) $\frac{2}{8} + 15\frac{1}{2} =$

24) $4\frac{9}{10} + 4\frac{1}{6} =$

25) $3\frac{3}{7} + 1\frac{3}{4} =$

26) $51 + 3\frac{5}{8} =$

27) $6\frac{3}{8} + 2\frac{2}{6} =$

Name: Date: Score: Unstoppable Owl

67) Adding Fractions (Mixed Review)

You've learned to add many types of fractions. Practice your skills on the problems below. Pay attention to whether or not the fractions have the same denominators.

Directions: *Add the mixed fractions and state your answer as a mixed fraction in simplified form.*

1) $\frac{5}{7} + \frac{1}{7} =$

2) $5\frac{1}{2} + 3\frac{5}{6} =$

3) $\frac{8}{9} + \frac{7}{9} =$

4) $6\frac{5}{8} + 7\frac{2}{3} =$

5) $5\frac{7}{10} + \frac{9}{10} =$

6) $3\frac{2}{5} + 6\frac{1}{3} =$

7) $2\frac{8}{12} + 57\frac{11}{12} =$

8) $\frac{4}{5} + 8\frac{4}{7} =$

9) $6\frac{1}{5} + \frac{4}{5} =$

10) $1\frac{7}{8} + 5\frac{1}{6} =$

11) $\frac{5}{9} + \frac{2}{3} =$

12) $2\frac{8}{12} + 5\frac{5}{6} =$

13) $\frac{4}{6} + \frac{14}{7} =$

14) $\frac{4}{10} + 11\frac{14}{20} =$

15) $8\frac{5}{6} + 6\frac{5}{6} =$

16) $\frac{9}{14} + \frac{13}{14} =$

17) $19\frac{1}{4} + \frac{5}{9} =$

18) $1\frac{1}{2} + 4\frac{2}{5} =$

19) $\frac{3}{8} + \frac{2}{3} =$

20) $2\frac{13}{15} + 25\frac{7}{15} =$

21) $4\frac{3}{6} + 3\frac{3}{5} =$

22) $\frac{4}{7} + \frac{5}{7} =$

23) $75\frac{2}{10} + \frac{1}{2} =$

24) $3\frac{7}{25} + 3\frac{18}{25} =$

25) $\frac{3}{4} + 47\frac{3}{4} =$

26) $1\frac{24}{40} + 5\frac{25}{40} =$

27) $\frac{3}{12} + \frac{2}{3} =$

28) $2\frac{5}{16} + 3\frac{9}{16} =$

29) $2\frac{7}{8} + 6\frac{6}{10} =$

30) $5 + 3\frac{2}{3} =$

31) $1\frac{1}{4} + 1\frac{2}{5} =$

32) $3\frac{1}{2} + 5\frac{13}{18} =$

33) $\frac{1}{6} + \frac{7}{8} =$

Name: Date: Score: Unstoppable Owl

68) Subtracting Fractions (With Unlike Denominators)

If you know how to add fractions with unlike denominators, then subtracting fractions with unlike denominators should be easy. Rewrite the fractions as equivalent fractions so that the denominators are the same. Once the denominators are the same, subtract the numerators. Then, simplify if necessary.

Examples:	Rewritten	Subtracted	Simplified Answer
$\frac{3}{4} - \frac{2}{6} =$	$\frac{9}{12} - \frac{4}{12} =$	$\frac{9}{12} - \frac{4}{12} = \frac{5}{12}$	$\frac{5}{12}$ is already simplified.
$\frac{4}{6} - \frac{1}{5} =$	$\frac{20}{30} - \frac{6}{30} =$	$\frac{20}{30} - \frac{6}{30} = \frac{14}{30}$	$\frac{14}{30} = \frac{7}{15}$

Directions: *Subtract each set of fractions and state your answer in simplified form.*

1) $\frac{3}{4} - \frac{5}{8} =$

2) $\frac{1}{2} - \frac{3}{14} =$

3) $\frac{2}{3} - \frac{1}{5} =$

4) $\frac{3}{4} - \frac{5}{7} =$

5) $\frac{4}{9} - \frac{1}{6} =$

6) $\frac{1}{2} - \frac{1}{5} =$

7) $\frac{9}{16} - \frac{1}{4} =$

8) $\frac{7}{8} - \frac{1}{2} =$

9) $\frac{5}{7} - \frac{3}{5} =$

10) $\frac{7}{25} - \frac{1}{5} =$

11) $\frac{3}{4} - \frac{3}{5} =$

12) $\frac{2}{3} - \frac{4}{9} =$

13) $\frac{5}{7} - \frac{1}{3} =$

14) $\frac{8}{10} - \frac{1}{2} =$

15) $\frac{4}{5} - \frac{3}{8} =$

16) $\frac{2}{3} - \frac{1}{6} =$

17) $\frac{11}{12} - \frac{2}{3} =$

18) $\frac{4}{5} - \frac{1}{7} =$

19) $\frac{7}{9} - \frac{2}{5} =$

20) $\frac{7}{8} - \frac{5}{6} =$

21) $\frac{7}{10} - \frac{3}{8} =$

22) $\frac{4}{7} - \frac{1}{4} =$

23) $\frac{3}{5} - \frac{1}{2} =$

24) $\frac{5}{6} - \frac{3}{5} =$

25) $\frac{5}{8} - \frac{1}{4} =$

26) $\frac{5}{9} - \frac{1}{6} =$

27) $\frac{11}{12} - \frac{3}{4} =$

Name: Date: Score: Unstoppable Owl

69) Subtracting Mixed Fractions (With Unlike Denominators)

Equivalent fractions can be used to create a common denominator when subtracting. Rewrite the fractions to create a common denominator and then subtract. If regrouping is needed to subtract the fractions, one whole can be regrouped from the whole number to create an improper fraction. (See page 62 for a review of subtracting mixed fractions.)

Examples:	**Rewritten & Regrouped**	**− Fractions**	**− Whole Numbers**	(Answer)
$7\frac{1}{2} - 2\frac{2}{6} =$	$7\frac{3}{6} - 2\frac{2}{6} =$ (No Regrouping Needed)	$\frac{3}{6} - \frac{2}{6} = \frac{1}{6}$	$7 - 2 = 5$	$5\frac{1}{6}$
$9\frac{2}{5} - 1\frac{2}{3} =$	$8\frac{21}{15} - 1\frac{10}{15} =$	$\frac{21}{15} - \frac{10}{15} = \frac{11}{15}$	$8 - 1 = 7$	$7\frac{11}{15}$

Directions: *Subtract. When possible, state your answer as a mixed fraction in simplified form.*

1) $8\frac{3}{4} - 6\frac{2}{5} =$

2) $6\frac{1}{2} - 1\frac{4}{7} =$

3) $3\frac{1}{3} - 2\frac{3}{4} =$

4) $7\frac{2}{3} - 2\frac{3}{5} =$

5) $14\frac{5}{9} - 9\frac{2}{3} =$

6) $4\frac{1}{6} - 1\frac{2}{5} =$

7) $9\frac{7}{10} - 4\frac{1}{2} =$

8) $5\frac{13}{20} - 4\frac{4}{5} =$

9) $8\frac{1}{3} - 1\frac{4}{12} =$

10) $9\frac{7}{15} - 7\frac{4}{5} =$

11) $7\frac{1}{2} - 5\frac{2}{3} =$

12) $7\frac{3}{5} - \frac{7}{8} =$

13) $6\frac{3}{6} - 3\frac{1}{2} =$

14) $17\frac{5}{8} - 2\frac{5}{6} =$

15) $3\frac{2}{3} - 1\frac{5}{9} =$

16) $8\frac{3}{10} - \frac{3}{4} =$

17) $9\frac{11}{16} - 5\frac{1}{2} =$

18) $6\frac{2}{4} - 3\frac{7}{20} =$

19) $7\frac{1}{7} - 4\frac{1}{3} =$

20) $8\frac{3}{10} - 5\frac{1}{2} =$

21) $6\frac{2}{3} - 5\frac{14}{15} =$

22) $45\frac{1}{6} - 7\frac{7}{10} =$

23) $6\frac{47}{75} - \frac{9}{25} =$

24) $7\frac{1}{4} - 2\frac{4}{9} =$

25) $6\frac{11}{25} - 4\frac{4}{10} =$

26) $5\frac{5}{18} - 2\frac{1}{2} =$

27) $10\frac{1}{4} - \frac{5}{7} =$

Name: Date: Score: Unstoppable Owl

70) Subtracting Fractions (Mixed Review)

Directions: *Subtract. When possible, state your answer as a mixed fraction in simplified form.*

1) $8\frac{4}{9} - 6\frac{7}{9} =$

2) $\frac{3}{5} - \frac{3}{7} =$

3) $2\frac{1}{4} - 1\frac{1}{2} =$

4) $7\frac{3}{8} - \frac{2}{3} =$

5) $13\frac{2}{5} - 8\frac{5}{6} =$

6) $2\frac{3}{4} - 1\frac{7}{12} =$

7) $\frac{11}{13} - \frac{8}{13} =$

8) $5 - 4\frac{3}{5} =$

9) $6\frac{17}{24} - 3\frac{3}{8} =$

10) $32\frac{5}{8} - 7\frac{4}{5} =$

11) $7\frac{1}{4} - 3\frac{2}{9} =$

12) $\frac{7}{8} - \frac{2}{3} =$

13) $9\frac{1}{2} - 2\frac{5}{12} =$

14) $5\frac{1}{6} - 4\frac{1}{3} =$

15) $4\frac{3}{5} - \frac{5}{6} =$

16) $9\frac{1}{6} - 3\frac{1}{4} =$

17) $\frac{1}{2} - \frac{4}{20} =$

18) $6\frac{3}{8} - 4\frac{7}{8} =$

19) $8\frac{11}{14} - \frac{1}{2} =$

20) $7\frac{1}{12} - 5\frac{7}{12} =$

21) $4\frac{1}{4} - 1\frac{5}{11} =$

22) $9\frac{1}{5} - 8\frac{3}{4} =$

23) $2\frac{2}{3} - 2\frac{1}{2} =$

24) $13 - 5\frac{3}{7} =$

25) $7\frac{3}{8} - 4\frac{1}{6} =$

26) $5\frac{5}{12} - 1\frac{5}{6} =$

27) $4\frac{1}{2} - 2\frac{11}{16} =$

28) $11\frac{1}{9} - 6\frac{1}{3} =$

29) $10\frac{3}{4} - \frac{7}{8} =$

30) $8\frac{3}{12} - 6\frac{1}{4} =$

31) $\frac{1}{2} - \frac{2}{9} =$

32) $7\frac{5}{18} - 3\frac{1}{2} =$

33) $1\frac{3}{5} - \frac{6}{9} =$

34) $9\frac{4}{10} - 6\frac{2}{3} =$

35) $7 - \frac{3}{8} =$

36) $4\frac{2}{4} - 3\frac{4}{5} =$

Name: Date: Score: Unstoppable Owl

71) Percentages (Converting Decimals to Percentages)

Remember how a number can be represented by a decimal or a fraction? Well, numbers can be written as a **percentage** too. The word **percent** means "out of 100" and is represented by the symbol %. 25% means 25 percent.

Percentages are like decimals and fractions. They're a different way to represent numbers. One fourth can be written as a fraction, as the decimal 0.25, or as 25%.

$$\frac{1}{4} = \frac{25}{100} = 0.25 = 25\%$$

Fraction Decimal Percent

Decimal:	(2 places to the right.)	Percentage:
0.25	0.25	25%
0.7	0.7	70%
4.19	4.19	419%
0.06	0.06	6%
0.005	0.005	0.5%

Converting Decimals to Percentages:

To convert decimals to percentages, you can imagine the decimal point moving two places to the right. Then write a percentage symbol at the end of the number.

Directions:
Write each decimal as a percentage.

1) 0.48
2) 0.31
3) 0.2
4) 0.09
5) 0.74
6) 2.63
7) 0.01
8) 5.08
9) 0.99
10) 0.002

11) 1.37
12) 0.56
13) 3.105
14) 6
15) 0.4
16) 0.007
17) 1.052
18) 0.03
19) 2.6
20) 0.914

21) 0.22
22) 0.8
23) 0.16
24) 7.03
25) 0.001
26) 0.405
27) 0.55
28) 1.003
29) 0.6
30) 0.04

31) 0.061
32) 8.25
33) 4
34) 6.001
35) 0.0045
36) 0.1
37) 0.0008
38) 12.5
39) 3.7
40) 0.2075

Name: Date: Score: Unstoppable Owl

72) Percentages (Converting Percentages to Decimals)

Decimals can be converted to percentages by moving the decimal point two place values to the right. The opposite can be done to convert a percentage to a decimal. With a percentage, move the decimal place two place values to the left.

Percentage:	(2 places to the left.)	Decimal:
25%	25.	0.25
40%	40.	0.4
307%	307.	3.07
8%	8.	0.08
0.1%	0.1	0.001

If a decimal point is not written with a percentage, it means the percentage is a whole number. The decimal point is to the right of the whole number, even though it is not written. For example, 25% is the same as 25.0, and 8% is the same as 8.0 percent. The decimal point is to the right of the whole number.

Directions:
Write each percentage as a decimal.

1) 83%
2) 9%
3) 30%
4) 475%
5) 0.7%
6) 200%
7) 5.8%
8) 106%
9) 0.05%
10) 11%
11) 7.6%
12) 720%
13) 55%
14) 0.4%
15) 150%
16) 10%
17) 3%
18) 60.5%
19) 8.14%
20) 90%
21) 0.39%
22) 8%
23) 71%
24) 111%
25) 5%
26) 0.09%
27) 4.2%
28) 50.4%
29) 2.95%
30) 260%
31) 0.8%
32) 6.7%
33) 45%
34) 37.1%
35) 1000%
36) 60%
37) 2.003%
38) 8.5%
39) 0.172%
40) 800%
41) 2500%
42) 4195%
43) 0.063%
44) 506.2%

Name: Date: Score: Unstoppable Owl

73) Percentages (Converting Fractions to Percentages)

Fractions can also be converted to percentages. But it's often easier to convert the fractions to decimals first. Once a fraction is converted to a decimal, it's not hard to convert that decimal to a percentage.

Fraction:	Decimal:	(2 places to the right.)	Percentage:
$\frac{3}{4}$	**0.75**	**0.75**	**75%**
$\frac{1}{20}$	**0.05**	**0.05**	**5%**
$\frac{9}{10}$	**0.9**	**0.9**	**90%**
$3\frac{1}{2}$	**3.50**	**3.5**	**350%**
$1\frac{1}{8}$	**1.125**	**1.125**	**112.5%**

Directions:
Write each fraction as a percentage.

Common fraction to decimal conversions can be found on pages 23-25.

A calculator may be used for less common conversions.

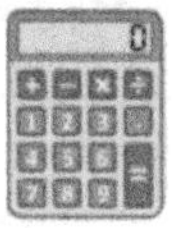

1) $\frac{3}{10}$

2) $\frac{1}{25}$

3) $2\frac{1}{4}$

4) $\frac{73}{100}$

5) $5\frac{9}{20}$

6) $\frac{2}{4}$

7) $\frac{7}{50}$

8) $1\frac{7}{10}$

9) $\frac{4}{5}$

10) $\frac{23}{25}$

11) $6\frac{1}{4}$

12) $\frac{19}{20}$

13) $\frac{3}{50}$

14) $20\frac{1}{2}$

15) $\frac{8}{10}$

16) $\frac{3}{20}$

17) $\frac{11}{25}$

18) $\frac{39}{100}$

19) $3\frac{6}{10}$

20) $\frac{7}{100}$

21) $\frac{11}{25}$

22) $\frac{9}{16}$

23) $4\frac{1}{20}$

24) $\frac{11}{80}$

25) $\frac{3}{5}$

26) $1\frac{1}{8}$

27) $\frac{19}{40}$

28) $2\frac{9}{50}$

29) $\frac{3}{8}$

30) $\frac{23}{400}$

31) $\frac{1}{40}$

32) $\frac{1}{200}$

33) $\frac{7}{1{,}000}$

34) $7\frac{3}{4}$

35) $\frac{1}{16}$

36) $\frac{7}{400}$

37) $3\frac{1}{2}$

38) $\frac{63}{1{,}000}$

39) $10\frac{1}{5}$

40) $\frac{361}{2{,}000}$

Name: Date: Score: *Unstoppable Owl*

74) Identifying Percentages

Use your knowledge of converting fractions and decimals to percentages to identify what percent of a shape is shaded. The first example has three fourths of the shape shaded, which is 0.75 when written as a decimal and 75% when written as a percentage.

If there is a repeating decimal, like example 2, abbreviate the decimal.

Examples: *What percent of the shape is shaded?*

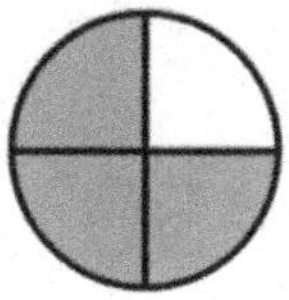 = 75%

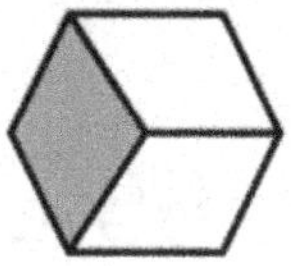 = $33.\overline{3}\%$

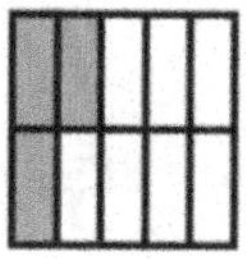 = 30%

Directions: *What percent of the shape is shaded?*

1) 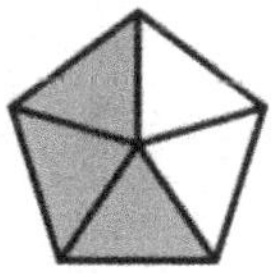 =

2) 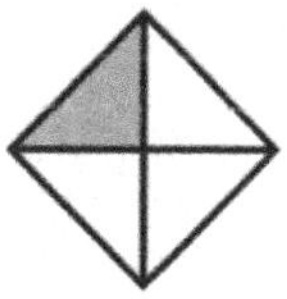 =

3) 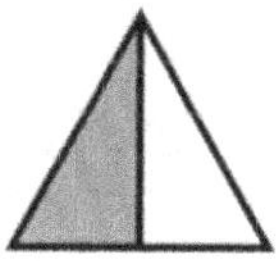 =

4) 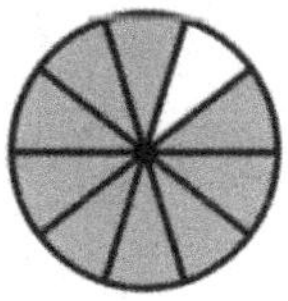 =

5)  =

6)  =

7)  =

8) 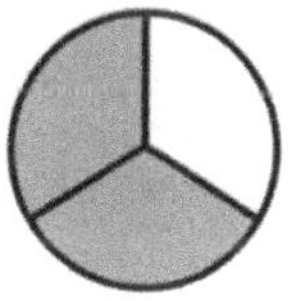 =

9) 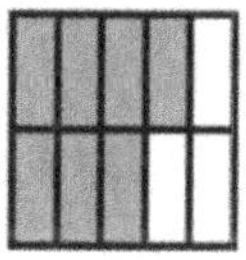 =

10) 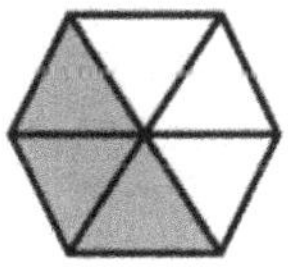 =

11)  =

12) 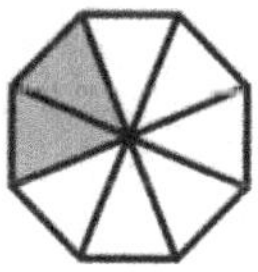 =

13)  =

14) 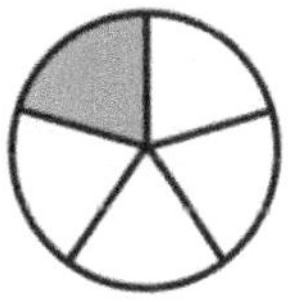 =

15)  =

16) 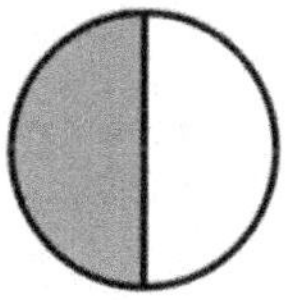 =

17) 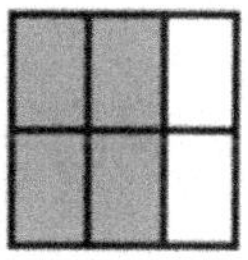 =

18) 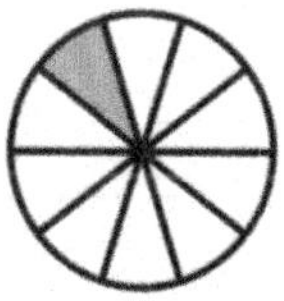 =

Name: Date: Score:

75) Identifying Percentages

Use your knowledge of percentages to shade each shape. Think about what each percentage represents when written as a decimal or as a fraction.

There may be many ways to correctly shade each shape. For instance, in the first example, the upper left space did not need to be shaded. Any one of the other spaces could have been shaded. If only one space is shaded, the answer will be correct.

Examples: *Shade the shape to match the percent.*

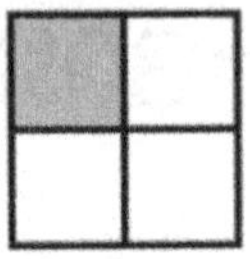 = 25%

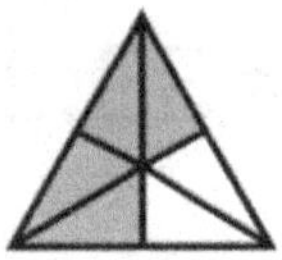 = $66.\overline{6}$%

 = 50%

Directions: *Shade the shape to match the percent.*

1) 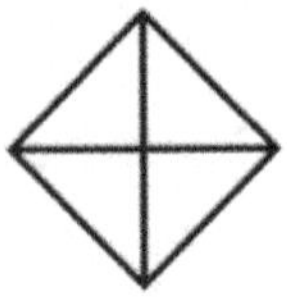 = 50%

2) 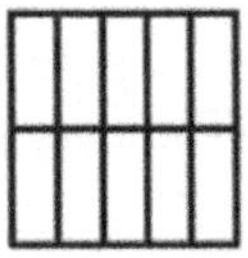 = 10%

3)  = $33.\overline{3}$%

4)  = 60%

5) 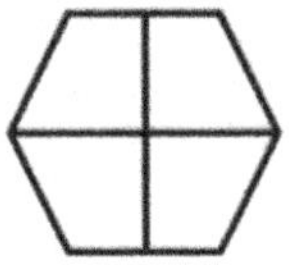 = 75%

6)  = 40%

7)  = 50%

8) 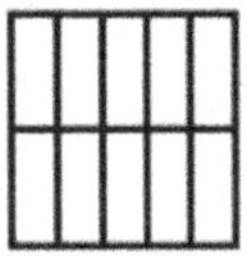 = 70%

9)  = 25%

10) 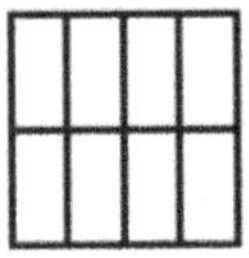 = 87.5%

11)  = 50%

12) 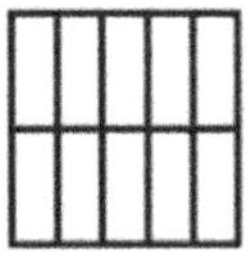 = 80%

13)  = 40%

14) 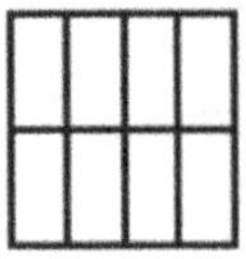 = 75%

15) 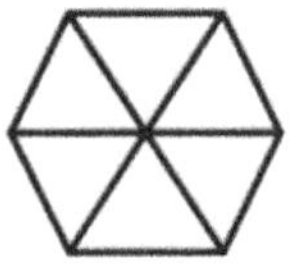 = 100%

16) 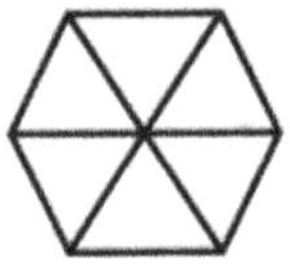 = $66.\overline{6}$%

17)  = 80%

18) 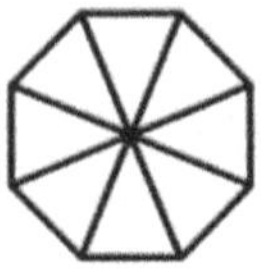 = 12.5%

Name: Date: Score: Unstoppable Owl

76) Identifying Percentages and Fractions

Percentages and fractions are often used to describe a portion of a group of items. For instance, someone may say that 75% of the flowers are yellow and 25% of the flowers are purple. They may also say that three fourths of the flowers are yellow and one fourth of the flowers are purple.

Being able to interpret what these percentages mean is important. Use the images on this page to help visualize and understand what the percentages mean. It may help to circle the correct number of objects.

Examples 1:

Megan picked 10 apples. If she gave 70% of them away, how many did she give away?

7 apples

Directions: *Use the images to answer each question.*

1) 50% of the 24 muffins are blueberry. How many blueberry muffins are there?

2) 25% of the 24 muffins are banana nut. How many banana nut muffins are there?

3) One third of the 30 marbles are green. How many green marbles are there?

4) One half of the 30 marbles are red. How many red marbles are there?

5) Alex had 30 marbles. Then, he lost 10% of them. How many marbles did Alex lose?

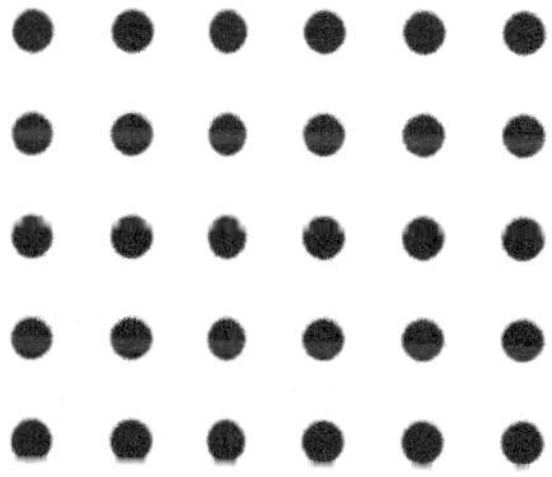

6) If 75% of the 20 flowers are yellow, how many flowers are yellow?

7) If 25% of the 20 flowers are purple, how many flowers are purple?

8) If Liz picked one fifth of the 20 flowers, how many flowers did she pick?

Name: Date: Score: Unstoppable Owl

77) Multiplying Decimals (With a calculator)

Pictures can help you understand how numbers relate to percentages and fractions, but pictures are not always available or easy to draw. It may not be difficult to figure out what 75% of 20 is. But it would be difficult to draw an example to figure out what 75% of 2,184 is.

Luckily, multiplication can be used to calculate the answers to problems that are difficult to figure out mentally or by using a drawing. To find 75% of 2,184, change the percentage to a decimal (75% = 0.75). Then, multiply (0.75 × 2,184 = 1,638).

The same is true when working with fractions. The fractions can be converted to a decimal before multiplying. So, if you needed to determine what two-fifths of 8,715 is, you could convert two-fifths to a decimal (0.4) and multiply (0.4 × 8,715 = 3,486).

Examples:	**Step 1:** Convert to a decimal.	**Step 2:** Multiply
What is 35% of 740?	35% = **0.35**	0.35 × 740 = **259**
What is $\frac{1}{20}$ of 6,120?	$\frac{1}{20}$ = **0.05**	0.05 × 6,120 = **306**

Directions: *Convert each fraction and percentage to a decimal. Then, use a calculator to multiply and find the answer to each question.*

1) What is 90% of 3,620?

2) What is $\frac{1}{4}$ of 1,068?

3) What is 56% of 300?

4) What is 8% of 500?

5) What is $\frac{1}{25}$ of 900?

6) What is 70% of 1,450?

7) What is 2% of 8,000?

8) What is $\frac{6}{50}$ of 750?

9) What is 150% of 400?

10) What is 200% of 600?

11) What is 12% of 825?

12) What is 1% of 500?

13) What is $\frac{3}{5}$ of 2,640?

14) What is 0.5% of 1,000?

15) What is 35% of 740?

16) What is $\frac{3}{4}$ of 10,000?

17) What is $\frac{3}{10}$ of 750?

18) What is 99% of 200?

19) What is 6% of 5,800?

20) What is $\frac{1}{25}$ of 2,000?

Name: Date: Score: Unstoppable Owl

78) Multiplying Decimals (With a Whole Number)

On the previous page, percentages and fractions were converted to decimals and then multiplied with other numbers. But a calculator is not always available to do this. It's good to learn how to multiply decimals with other numbers using a pencil and paper too.

The good news is, if you already know how to multiply multi-digit numbers, learning to multiply decimals is not hard. If you have not learned this skill, or would like a refresher, consider practicing with the book to the right. The book to the right is not required. Even if you have never multiplied multi-digit numbers, you may be able to study the examples below and learn the steps. It will just take some extra effort.

How to Multiply Decimals

When multiplying decimals, you must place the decimal point in your answer. Placing the decimal point in the answer is easy. First, you ignore the decimal points and multiply, as if you were pretending each number was a whole number. At the end, you count how many digits are to the right of the decimal point in the numbers that were multiplied. For instance, if there were a total of 2 numbers to the right of the decimal point before multiplying, there will be 2 digits to the right of the decimal point after multiplying. (*The only exception is if the furthest digit to the right is a zero. This still counts as a placeholder but the trailing zero does not need to be written in the final answer*.)

	Ignore the decimal point during these steps.				
Examples:	**Multiply**	**Multiply**	**Multiply**	**Add**	**Place the Decimal**
7.3 × 5 = (*1 decimal place*)	73 × 5 **15** (3 × 5 = **15**)	73 × 5 15 **350** (70 × 5 = **350**)		73 × 5 15 +350 365	**36.5** (*1 decimal place*)
2.16 × 7 = (*2 decimal places*)	216 × 7 **42** (6 × 7 = **42**)	216 × 7 42 **70** (10 × 7 = **70**)	216 × 7 42 70 **1,400** (200 × 7 = **1,400**)	216 × 7 42 70 +1,400 1,512	**15.12** (*2 decimal places*)

Directions: *Multiply each set of numbers. Remember to place the decimal in your answer.*

1) **3.9 × 8 =** 2) **25 × 1.75 =** 3) **0.4 × 55 =** 4) **6.1 × 3 =**

Name: Date: Score: Unstoppable Owl

79) Multiplying Decimals (With a Whole Number)

Directions: *Multiply each set of numbers. Remember to place the decimal in your answer.*

1) $5.3 \times 2 =$

2) $12 \times 0.68 =$

3) $1.85 \times 30 =$

4) $4 \times 3.2 =$

5) $45 \times 2.1 =$

6) $6 \times 7.2 =$

7) $0.47 \times 8 =$

8) $5.19 \times 2 =$

9) $2.8 \times 300 =$

10) $27 \times 4.5 =$

11) $9.4 \times 5 =$

12) $20 \times 0.15 =$

13) $38 \times 1.6 =$

14) $2.68 \times 7 =$

15) $0.03 \times 80 =$

16) $5.7 \times 4 =$

17) $3.25 \times 2 =$

18) $5.5 \times 3 =$

19) $6 \times 4.1 =$

20) $89 \times 0.84 =$

80) Multiplying Decimals

Multiplying two numbers that each have a decimal uses the same steps as multiplying a decimal and a whole number. The only difference is that when placing the decimal in the answer, you need to count the digits to the right of the decimal point in each of the numbers.

For instance, if the first number has 2 digits to the right of the decimal point and the second number has 1 digit to the right of the decimal point, you will need to add all these digits together to determine where the place the decimal point in the answer. In this case, 1 digit plus 2 digits = 3 digits. So, the answer will have 3 digits to the right of the decimal point.

	Ignore the decimal point during these steps.						
Example:	**Multiply**	**Multiply**	**Multiply**	**Multiply**	**Multiply**	**Multiply and Add**	**Place the Decimal**
2.71 × 3.5 = *(3 decimal places)*	271 × 35 **5**	271 × 35 5 **350**	271 × 35 5 350 **1,000**	271 × 35 5 350 1,000 **30**	271 × 35 5 350 1,000 30 **2,100**	271 × 35 5 350 1,000 30 2,100 **+6,000** 9,485	**9.485** *(3 decimal places)*

Directions: *Multiply each set of numbers. Remember to place the decimal in your answer.*

1) **5.3 × 2.7 =**

2) **3.61 × 0.9 =**

3) **4.5 × 7.38 =**

4) **0.6 × 0.2 =**

5) **1.563 × 2.4 =**

6) **4.06 × 1.8 =**

7) **0.22 × 2.17 =**

8) **3.9 × 6.8 =**

Name: Date: Score: Unstoppable Owl

81) Multiplying Decimals

Directions: *Multiply each set of numbers. Remember to place the decimal in your answer.*

1) **0.35 × 4.1 =**

2) **7.1 × 8.3 =**

3) **2.01 × 4.14 =**

4) **3.2 × 7.9 =**

5) **4.5 × 2.109 =**

6) **3.16 × 0.28 =**

7) **5.21 × 2.7 =**

8) **6.41 × 2.06 =**

9) **2.8 × 3.08 =**

10) **4.7 × 1.5 =**

11) **9.14 × 2.25 =**

12) **2.003 × 1.65 =**

13) **6.7 × 0.3 =**

14) **1.72 × 0.04 =**

15) **3.62 × 1.8 =**

16) **4.7 × 1.181 =**

Name: Date: Score: Unstoppable Owl

82) Multiplying Decimals

Once you know how to multiply decimals, you can solve a lot of problems—even problems that involve percentages or fractions. Look back to page 77. On this page, percentages and fractions were converted to decimals. Then, these decimals were multiplied with other numbers.

On this page, you are using the same skills. The only difference is that calculators are no longer allowed. All the skills needed to solve these problems without a calculator have been practice. All you need to do is combine these skills to solve each problem.

Directions: *Convert each fraction and percentage to a decimal. Then, use multiplication to answer each question. (Calculators are not allowed.)*

1) **What is 70% of 400?**

2) **What is $\frac{1}{5}$ of 28?**

3) **What is 32% of 80?**

4) **What is $\frac{1}{4}$ of 7?**

5) **What is $\frac{6}{25}$ of 900?**

6) **What is 15% of 60?**

7) **What is $\frac{3}{4}$ of 200?**

8) **What is $\frac{11}{20}$ of 9?**

9) **What is 3% of 2?**

10) **What is 35% of 740?**

83) Dividing Decimals (With a Whole Number)

Dividing with decimals is very similar to dividing with multi-digit numbers. Like multiplying decimals, most of the steps can be completed while ignoring the decimal points. Only at the end do the decimal points need to be taken into consideration. Count the total number of place values to the right of the decimal and insert that many place values to the right of the decimal in your answer.

If you are not proficient with multi-digit division, the book to the right can be used to develop this skill. Dividing large numbers is not a focus of this book. You can still study the examples below and try to learn the steps. It will just take more time and effort.

The examples below use the **long division method**. It's often used when dividing with decimals. You may use other methods (such as **partial products method**) if you know them.

Example 1: $4.2 \div 3 =$	**Example 2:** $1.74 \div 2 =$
$3\overline{)42}$ quotient **14** -3 ← $(3 \times \mathbf{1} = 3)$ 12 -12 ← $(3 \times \mathbf{4} = 12)$ **0**	$2\overline{)174}$ quotient **87** -16 ← $(2 \times \mathbf{8} = 16)$ 14 -14 ← $(2 \times \mathbf{7} = 14)$ **0**
Place the decimal: **1.4** (1 decimal place)	**Place the decimal:** **0.87** (2 decimal places)
Answer: $4.2 \div 3 =$ **1.4**	**Answer:** $1.74 \div 2 =$ **0.87**

Directions: *Divide each set of numbers. Remember to place the decimal in your answer.*

1) **7.5 ÷ 5 =**

2) **0.8 ÷ 4 =**

3) **8.26 ÷ 2 =**

4) **0.72 ÷ 3 =**

5) **19.2 ÷ 8 =**

6) **11.8 ÷ 2 =**

7) **2.45 ÷ 7 =**

8) **5.04 ÷ 4 =**

Name: Date: Score: Unstoppable Owl

84) Dividing Decimals (With a Whole Number)

On the previous page, the original numbers and the answer had the same amount of place values to the right of the decimal point. This doesn't always happen though. Sometimes, even when you are ignoring the decimal places and dividing, you will end up with a decimal. When that happens, you must move the decimal place in the final answer. The decimal point should be moved to the left one place value for each of the place values that are to the right of the decimal points in the original numbers.

For instance, if there were 2 place values to the right of the decimal in the original numbers, and there was one number to the right of the decimal while dividing and ignoring the decimal point, there would be a total of 3 decimal places to the right of the decimal in the final answer.

Example 1: 2.1 ÷ 4 =	**Example 2:** 0.75 ÷ 6 =
5.25 4) 21.00 − 20 ← (4 × **5** = 20) 10 − 8 ← (4 × **2** = 8) 20 − 20 ← (4 × **5** = 20) 0	**12.5** 6) 75.0 − 6 ← (6 × **1** = 6) 15 − 12 ← (6 × **2** = 12) 30 − 30 ← (6 × **5** = 30) 0
Place the decimal: **0.525** (1 decimal place)	**Place the decimal:** **0.125** (2 decimal places)
Answer: 2.1 ÷ 4 = **0.525**	**Answer:** 0.75 ÷ 6 = **0.125**

Directions: *Divide each set of numbers. Remember to place the decimal in your answer.*

1) **8.3 ÷ 5 =** 2) **0.07 ÷ 2 =** 3) **4.6 ÷ 8 =** 4) **25.3 ÷ 2 =**

5) **0.87 ÷ 4 =** 6) **0.6 ÷ 5 =** 7) **7.58 ÷ 4 =** 8) **4.28 ÷ 8 =**

Name: Date: Score: Unstoppable Owl

85) Dividing Decimals (With a Whole Number)

Directions: *Divide each set of numbers. Remember to place the decimal in your answer.*

1) **1.65 ÷ 3 =**

2) **0.3 ÷ 4 =**

3) **7.16 ÷ 4 =**

4) **3.678 ÷ 2 =**

5) **6.93 ÷ 7 =**

6) **32.9 ÷ 2 =**

7) **78.3 ÷ 9 =**

8) **0.58 ÷ 8 =**

9) **40.5 ÷ 9 =**

10) **3.9 ÷ 4 =**

11) **0.03 ÷ 5 =**

12) **4.32 ÷ 6 =**

13) **3.78 ÷ 4 =**

14) **5.49 ÷ 9 =**

15) **37.4 ÷ 8 =**

16) **8.12 ÷ 7 =**

Name: Date: Score: Unstoppable Owl

86) Dividing Decimals (With a Whole Number)

When dividing, the answer may be a decimal that is very long. When this happens, the answer may be rounded. On this page, round long answers to **the nearest hundredth**.

The examples below show how to round the answer to the nearest hundredth. They do not show how to perform the division. Those skills were covered in previous pages.

When dividing and rounding to the nearest hundredth, *you only need to divide long enough to find the value of the thousandth place*. The thousandth place is needed to know whether to round the hundredths place up or down.

Examples:	**Answers:**	**Rounded Answers:** (*to the nearest hundredth*)
0.5 ÷ 8 =	**0.0625**	**0.06**
0.7 ÷ 3 =	**0.23333....**	**0.23**
8.1 ÷ 7 =	**1.157142**	**1.16**
20.92 ÷ 6 =	**3.486666...**	**3.49**
1.86 ÷ 9 =	**0.206666...**	**0.21**
9.37 ÷ 8 =	**1.17125**	**1.17**

Directions: *Divide each set of numbers and* ***round the answers to the nearest hundredth****.*

1) **0.7 ÷ 6 =**
2) **4.05 ÷ 8 =**
3) **42.1 ÷ 9 =**
4) **5.95 ÷ 4 =**

5) **83.6 ÷ 3 =**
6) **0.88 ÷ 6 =**
7) **7.15 ÷ 8 =**
8) **2.9 ÷ 6 =**

Name: Date: Score: Unstoppable Owl

87) Multiplying Fractions

When multiplying fractions, it helps to convert any whole numbers and mixed fractions into improper fractions. After that is done, multiply the numerators. Then, Multiply the denominators. The fraction that is created may then be reduced to its simplest form and converted to a mixed fraction when possible.

Step 1: Convert whole numbers and mixed fractions into an improper fraction.
Step 2: Multiply the numerators. Then, multiply the denominators.
Step 3: Convert to a mixed fraction and reduce the fraction (when possible).

Examples:	Step 1:	Step 2:	Step 3:
$\frac{5}{8} \times \frac{2}{3} =$	None	$\frac{5 \times 2}{8 \times 3} = \mathbf{\frac{10}{24}}$	$\frac{10}{24} = \mathbf{\frac{5}{12}}$
$4\frac{1}{2} \times \frac{3}{5} =$	$\frac{9}{2} \times \frac{3}{5} =$	$\frac{9 \times 3}{2 \times 5} = \mathbf{\frac{27}{10}}$	$\frac{27}{10} = 2\mathbf{\frac{7}{10}}$
$2 \times 3\frac{1}{4} =$	$\frac{2}{1} \times \frac{13}{4} =$	$\frac{2 \times 13}{1 \times 4} = \mathbf{\frac{26}{4}}$	$\frac{26}{4} = \frac{13}{2} = 6\mathbf{\frac{1}{2}}$

Directions: *Multiply. When possible, state each answer as a mixed fraction in simplest form.*

1) $\frac{1}{3} \times 2\frac{4}{5} =$

2) $1\frac{3}{4} \times 2\frac{2}{3} =$

3) $10 \times 3\frac{2}{5} =$

4) $\frac{5}{8} \times 2\frac{1}{2} =$

5) $1\frac{3}{5} \times \frac{4}{10} =$

6) $\frac{5}{6} \times \frac{2}{3} =$

7) $1\frac{2}{6} \times 2\frac{3}{4} =$

8) $1\frac{3}{7} \times 4\frac{1}{2} =$

9) $2\frac{2}{3} \times 1 =$

10) $\frac{5}{6} \times \frac{5}{8} =$

11) $2\frac{1}{4} \times 2\frac{1}{3} =$

12) $\frac{1}{5} \times 12\frac{1}{2} =$

13) $5 \times 1\frac{5}{6} =$

14) $4\frac{1}{2} \times \frac{7}{9} =$

Name: Date: Score: Unstoppable Owl

88) Multiplying Fractions

Directions: *Multiply. When possible, state each answer as a mixed fraction in simplest form.*

1) $1\frac{3}{5} \times 2\frac{2}{3} =$

2) $1\frac{1}{5} \times 1\frac{1}{5} =$

3) $\frac{1}{4} \times \frac{1}{3} =$

4) $3\frac{2}{6} \times \frac{3}{10} =$

5) $4\frac{2}{5} \times 1\frac{2}{5} =$

6) $\frac{6}{7} \times 2\frac{1}{2} =$

7) $7\frac{1}{2} \times 4 =$

8) $\frac{3}{8} \times \frac{8}{9} =$

9) $5\frac{1}{3} \times \frac{1}{6} =$

10) $\frac{3}{4} \times 2\frac{1}{3} =$

11) $4 \times 1\frac{5}{6} =$

12) $10 \times 2\frac{1}{2} =$

13) $4\frac{1}{2} \times 3\frac{1}{2} =$

14) $8\frac{1}{2} \times \frac{1}{2} =$

15) $\frac{1}{8} \times 7\frac{1}{3} =$

16) $1\frac{3}{4} \times 2\frac{1}{2} =$

17) $\frac{5}{6} \times \frac{4}{5} =$

18) $4\frac{2}{3} \times \frac{2}{7} =$

19) $3\frac{1}{2} \times 1\frac{4}{5} =$

20) $5 \times 2\frac{2}{3} =$

21) $2\frac{1}{2} \times 2\frac{1}{2} =$

22) $\frac{8}{10} \times \frac{3}{4} =$

23) $4\frac{2}{7} \times \frac{1}{6} =$

24) $1\frac{4}{5} \times 8 =$

Name: Date: Score: Unstoppable Owl

89) Multiplying Fractions

Directions: *Multiply. When possible, state each answer as a mixed fraction in simplest form.*

1) $\frac{1}{9} \times 1\frac{2}{3} =$

2) $7 \times 1\frac{1}{3} =$

3) $10 \times 3\frac{3}{5} =$

4) $2\frac{1}{3} \times 3\frac{1}{2} =$

5) $9\frac{1}{5} \times \frac{1}{2} =$

6) $\frac{6}{7} \times 2\frac{1}{3} =$

7) $2\frac{3}{4} \times 1\frac{2}{3} =$

8) $\frac{5}{8} \times \frac{9}{10} =$

9) $2\frac{2}{5} \times \frac{3}{4} =$

10) $3\frac{1}{4} \times 8\frac{2}{3} =$

11) $6 \times 2\frac{1}{3} =$

12) $1\frac{1}{4} \times 1\frac{1}{4} =$

13) $\frac{5}{6} \times \frac{3}{10} =$

14) $2\frac{2}{3} \times \frac{3}{5} =$

15) $\frac{1}{7} \times 9 =$

16) $1\frac{4}{5} \times 4\frac{1}{2} =$

17) $2\frac{1}{4} \times 3\frac{1}{3} =$

18) $\frac{4}{6} \times \frac{3}{4} =$

19) $3\frac{4}{7} \times \frac{3}{10} =$

20) $1\frac{2}{5} \times 10\frac{1}{2} =$

21) $\frac{5}{8} \times 1\frac{2}{3} =$

22) $7\frac{1}{2} \times 3 =$

23) $\frac{1}{8} \times \frac{1}{9} =$

24) $5\frac{1}{4} \times \frac{1}{8} =$

Name: Date: Score: Unstoppable Owl

90) Dividing Fractions

Dividing fractions is similar to multiplying fractions. In fact, you can rewrite the problem and use multiplication to calculate the answer. To rewrite the problem, take the number that is being divided by and ***flip** its numerator and denominator*. After that, multiply the fractions.

Step 1: Convert whole numbers and mixed fractions into an improper fraction.
Step 2: Make a multiplication problem by rewriting the fraction (flip the fraction).
Step 3: Multiply the numerators. Then, multiply the denominators.
Step 4: Convert to a mixed fraction and reduce the fraction (when possible).

Examples:	Step 1:	Step 2 and Step 3:	Step 4:
$\frac{4}{5} \div \frac{1}{3} =$	None	$\frac{4 \times 3}{5 \times 1} = \mathbf{\frac{12}{5}}$	$\frac{12}{5} = 2\frac{2}{5}$
$2\frac{1}{3} \div \frac{1}{4} =$	$\frac{7}{3} \times \frac{1}{4} =$	$\frac{7 \times 4}{3 \times 1} = \mathbf{\frac{28}{3}}$	$\frac{28}{3} = 9\frac{1}{3}$
$6 \div 1\frac{1}{8} =$	$\frac{6}{1} \times \frac{9}{8} =$	$\frac{6 \times 8}{1 \times 9} = \mathbf{\frac{48}{9}}$	$\frac{48}{9} = \frac{16}{3} = 5\frac{1}{3}$

Directions: *Divide. When possible, state each answer as a mixed fraction in simplest form.*

1) $\frac{4}{5} \div 1\frac{1}{3} =$

2) $1\frac{3}{7} \div 2\frac{1}{2} =$

3) $10 \div 3\frac{1}{2} =$

4) $\frac{1}{2} \div 5\frac{5}{6} =$

5) $3\frac{2}{6} \div \frac{3}{4} =$

6) $\frac{4}{5} \div \frac{5}{8} =$

7) $3\frac{1}{3} \div 2\frac{1}{4} =$

8) $1\frac{3}{5} \div 1\frac{1}{6} =$

9) $1\frac{1}{4} \div 8 =$

10) $\frac{5}{6} \div \frac{1}{8} =$

11) $5\frac{1}{2} \div 2\frac{2}{3} =$

12) $\frac{7}{8} \div 1\frac{2}{5} =$

13) $9 \div 3\frac{1}{3} =$

14) $2\frac{1}{2} \div \frac{3}{4} =$

Name: Date: Score: Unstoppable Owl

91) Dividing Fractions

Directions: *Divide. When possible, state each answer as a mixed fraction in simplest form.*

1) $\frac{1}{6} \div \frac{2}{3} =$

2) $1\frac{1}{4} \div \frac{6}{7} =$

3) $1\frac{3}{4} \div 4 =$

4) $2\frac{1}{3} \div 3\frac{1}{3} =$

5) $7\frac{1}{2} \div 1\frac{1}{4} =$

6) $\frac{7}{8} \div \frac{1}{3} =$

7) $2\frac{1}{3} \div \frac{5}{9} =$

8) $4\frac{1}{2} \div \frac{2}{5} =$

9) $\frac{8}{9} \div 6 =$

10) $6\frac{2}{3} \div 5 =$

11) $\frac{2}{3} \div \frac{1}{9} =$

12) $1\frac{3}{4} \div \frac{3}{8} =$

13) $\frac{7}{8} \div 1\frac{1}{10} =$

14) $9 \div 1\frac{2}{5} =$

15) $3 \div 1\frac{1}{2} =$

16) $2\frac{1}{4} \div 1\frac{3}{5} =$

17) $8 \div \frac{1}{2} =$

18) $\frac{3}{4} \div 9 =$

19) $2\frac{1}{3} \div 4\frac{1}{2} =$

20) $\frac{1}{5} \div \frac{2}{3} =$

21) $3\frac{2}{3} \div \frac{1}{3} =$

22) $1\frac{3}{4} \div 2\frac{2}{3} =$

23) $5 \div 2\frac{3}{4} =$

24) $6\frac{2}{3} \div 1\frac{1}{4} =$

Name: Date: Score: Unstoppable Owl

92) Dividing Fractions

Directions: *Divide. When possible, state each answer as a mixed fraction in simplest form.*

1) $3\frac{1}{2} \div \frac{1}{2} =$

2) $8\frac{3}{4} \div \frac{1}{2} =$

3) $\frac{9}{10} \div 2\frac{2}{3} =$

4) $1\frac{5}{6} \div 1\frac{1}{4} =$

5) $\frac{3}{20} \div \frac{1}{50} =$

6) $1\frac{5}{6} \div \frac{2}{3} =$

7) $2 \div 4\frac{1}{2} =$

8) $8 \div 2\frac{2}{9} =$

9) $5\frac{1}{2} \div 1\frac{2}{3} =$

10) $\frac{3}{10} \div \frac{8}{25} =$

11) $6\frac{2}{8} \div \frac{1}{2} =$

12) $1\frac{4}{5} \div 10 =$

13) $1\frac{1}{6} \div 2\frac{1}{2} =$

14) $\frac{4}{9} \div 1\frac{1}{4} =$

15) $\frac{1}{12} \div \frac{3}{4} =$

16) $7\frac{1}{2} \div \frac{5}{6} =$

17) $\frac{2}{5} \div \frac{2}{5} =$

18) $\frac{1}{6} \div 3\frac{1}{2} =$

19) $5\frac{2}{4} \div 2 =$

20) $\frac{5}{7} \div \frac{3}{8} =$

21) $1\frac{3}{5} \div \frac{1}{4} =$

22) $\frac{1}{3} \div 2\frac{1}{3} =$

23) $6 \div 1\frac{4}{5} =$

24) $12 \div \frac{1}{4} =$

Name: Date: Score: Unstoppable Owl

93) Mixed Review (Decimals)

Directions: *Solve each problem.*

1) **1.8 + 3.59 =**

2) **72.5 + 87.6 =**

3) **6.407 + 4.294 =**

4) **0.48 + 0.852 =**

5) **9 – 2.63 =**

6) **5.6 – 3.831 =**

7) **26.84 – 8.9 =**

8) **0.73 – 0.065 =**

9) **5.8 × 4.07 =**

10) **0.2 × 6.1 =**

11) **3.14 × 5.2 =**

12) **2.03 × 4.25 =**

13) **3.32 ÷ 4 =**

14) **4.7 ÷ 5 =**

15) **2.08 ÷ 8 =**

16) **5.7 ÷ 4 =**

Name: Date: Score: Unstoppable Owl

94) Mixed Review (Fractions)

Directions: *Solve each problem. State your answer as a mixed fraction in simplest form.*

1) $2\frac{3}{7} + 6\frac{1}{2} =$

2) $1\frac{2}{5} + 5 =$

3) $1\frac{5}{6} + 3\frac{2}{3} =$

4) $4\frac{3}{4} + 1\frac{2}{3} =$

5) $\frac{3}{5} + 3\frac{1}{8} =$

6) $\frac{1}{2} + \frac{7}{9} =$

7) $3\frac{1}{2} + 3\frac{5}{8} =$

8) $2\frac{9}{10} + \frac{2}{3} =$

9) $6\frac{1}{4} + \frac{2}{5} =$

10) $4\frac{1}{2} - 1\frac{3}{5} =$

11) $5\frac{1}{3} - 3 =$

12) $6\frac{3}{5} - 3\frac{1}{4} =$

13) $2\frac{1}{3} - 1\frac{1}{2} =$

14) $7\frac{2}{3} - 3\frac{3}{4} =$

15) $9\frac{1}{3} - 1\frac{3}{8} =$

16) $6\frac{4}{5} - 6\frac{1}{6} =$

17) $8\frac{2}{3} - \frac{5}{9} =$

18) $1\frac{1}{4} - \frac{5}{6} =$

19) $\frac{2}{3} \times 2\frac{4}{5} =$

20) $\frac{3}{7} \times 1\frac{1}{6} =$

21) $1\frac{1}{4} \times 2\frac{1}{3} =$

22) $2\frac{2}{3} \times 5 =$

23) $11 \times 3\frac{1}{2} =$

24) $\frac{2}{5} \times \frac{5}{8} =$

25) $\frac{5}{9} \div 2\frac{1}{2} =$

26) $3\frac{1}{4} \div \frac{1}{2} =$

27) $1\frac{3}{4} \div \frac{3}{10} =$

28) $\frac{1}{5} \div 1\frac{1}{3} =$

29) $\frac{5}{6} \div \frac{2}{9} =$

30) $5 \div 6\frac{1}{4} =$

Unstoppable Owl

Name: Date: Score:

95) Mixed Review (Percentages)

Directions: *Write each decimal as a percentage.*

1) **0.78** 2) **2.59** 3) **0.12** 4) **0.07**

5) **0.64** 6) **0.002** 7) **0.9** 8) **1.4**

9) **0.3** 10) **4.705** 11) **0.016** 12) **3.08**

Directions: *Write each percentage as a decimal.*

13) **27%** 14) **3%** 15) **185%** 16) **0.7%**

17) **5.6%** 18) **400%** 19) **70.4%** 20) **1,500%**

21) **0.61%** 22) **2.8%** 23) **0.06%** 24) **5%**

Directions: *What percent of the shape is shaded?*

25) 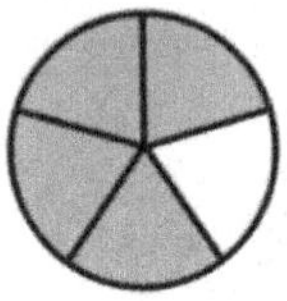 =

26)  =

27) 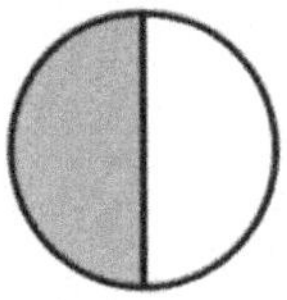 =

28)  =

29) 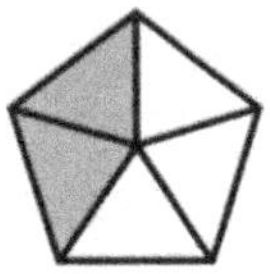 =

30) 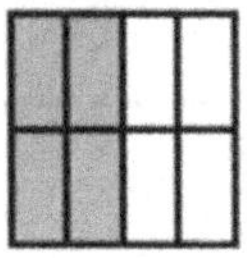 =

Directions: *Use your knowledge of percentages and multiplying decimals to solve each problem.*

31) **What is 70% of 200?**

32) **What is 5% of 80?**

33) **What is 150% of 60?**

34) **What is 300% of 25?**

35) **What is 62% of 500?**

36) 40% of the 90 flowers are purple. How many purple flowers are there?

37) 75% of the 300 fish are yellow. How many yellow fish are there?

38) There are 50 trees. 10% of them are oak trees. How many oak trees are there?

39) 4% of the 13,000 players will win prizes. How many players will win prizes?

40) The park had 28 swans. Then, the number of swans increased by 250%. How many swans are there now?

Name: Date: Score: Unstoppable Owl

96) Word Problems

1) Ben and Lily started a new workout routine. The first day, Lily recorded 13,000 steps on her fitness watch. Ben recorded four-fifths of the steps Lily did. How many steps did Ben record?

2) Ben's step goal for the second day was 12,000 steps, but by the end of the day he had passed his goal by 23%. How many steps did Ben get this day?

3) Lily's goal the second day was 12,000 steps. But she ended the day with 15% fewer steps than her goal. How many steps did Lily have the second day?

4) Ben and Lily are hiking a trail that is 7.2 km long. If they've already hiked 5.38 km of the trail, how much further do they need to hike to reach the end of the trail?

5) At the start of the new workout routine, Ben could do 18 push-ups. Two months later, the number of push-ups Ben could do increased by 50%. How many push-ups is that?

6) Lily could do 6 push-ups at the start of the new workout routine. Two months later, she could do 15 push-ups. How many percentage points did Lily's push-ups increase by.

7) Lily did yoga for 45 minutes.
Ben did yoga for a third of that time. How many minutes did Ben do yoga?

8) Ben went on a 52-minute bike ride. Lily went on a bike ride for three-fourths of the time Ben did. How many minutes long was Lily's bike ride?

Name: Date: Score:

Unstoppable Owl

97) Word Problems

1) **Joan's restaurant had 1,630 customers on Friday. If Joan expects to have 20% more customers on Saturday, how many customers is she expecting on Saturday?**

2) **The restaurant sold 580 salads, a fourth of which were Caesar salads. How many Caesar salads were sold?**

3) **The restaurant has 270 small plates and three and a half times as many large plates. How many large plates does the restaurant have?**

4) **The restaurant had 1,100 cups but over time, 17% of them were broken and thrown away. How many cups does the restaurant have left?**

5) **The restaurant served 1,430 cups of beverages today. Two-fifths of the beverages were cups of coffee. How many cups of coffee were served today?**

6) **The restaurant can seat 200 customers at a time. There are currently 140 customers seated in the restaurant. What percentage of the seats are being used?**

7) **The restaurant's soup recipe requires two and three-fourths bags of noodles for each batch of soup. How many bags of noodles are needed to make 3 batches of soup?** *State your answer as a mixed fraction in simplest form.*

8) **There were 40 boxes of napkins. Now there are only 16 boxes left. In simplest terms, what fraction of the 40 napkin boxes are left?**

Name: Date: Score: Unstoppable Owl

98) Word Problems

1) **Mike's construction business is building a house. The workers had 2 full bins of nails. Now, one bin is a third full and the other is three-fourths full. How many bins of nails are left?** *State your answer as a mixed fraction in simplest form.*

2) **It takes two-thirds of a box of nails to fill a nail gun with nails. If there are 11 nail guns to fill, how many boxes of nails are needed to fill them all?** *State your answer as a mixed fraction in simplest form.*

3) **A nail gun can hold 400 nails at a time. Mike's nail gun was full of nails but he's since used 38% of them. How many nails has he used?**

4) **One group of workers was given 4 and a half boxes of screws. The other group of workers was given 3 and three-fourths boxes of screws. How many boxes of screws is that altogether?** *State your answer as a mixed fraction in simplest form.*

5) **Each brick is 19.37 cm long. If 6 of these bricks are lined up lengthwise, how long will the row of bricks be?**

6) **Mike has a board that is 94.2 cm long. If he cuts 17.5 cm off this board, how long will the remaining board be?**

7) **A load of cement is being poured. The first eighth of the load filled five and a fourth sections of sidewalk. At this rate, how many sections of sidewalk will the entire load of cement fill?**

8) **Full cure time for the cement is 28 days. The cement has already been curing for 8 days. In simplest terms, what fraction of the full cure time is that?**

Name: Date: Score: Unstoppable Owl

99) Word Problems

1) Sue's goal was to finish the race in 50 seconds. She finished the race in 49.28 seconds. How many seconds did she beat her goal by?

2) Tom finished the race in 46.17 seconds. Kyle finished the race in 52.3 seconds. How much faster did Tom finish the race than Kyle?

3) Each person on a 4-person relay team had an average run time of 28.3 seconds. What was the team's total time for the race?

4) A different 4-person relay team had a total time of 121.64 seconds for the race. What was the average run time for each of the 4 people on the relay team?

5) Of the 80 people running the race, 6 were disqualified for false starting. What percentage of the runners were disqualified for false starting?

6) Of the 80 racers, the fastest 16 racers will receive a prize. In simplest form, what fraction of the racers will receive a prize?

7) The first-place winner finished the race in 39.05 seconds. The world record is 34.6 seconds. How much faster is the world record than the first-place winner's time?

8) Today, Luke finished the race in 43.11 seconds. This was not his fastest time though. Luke's fastest time was one-fourth of a second faster than today's time. What is Luke's fastest time?

Name: Date: Score:

Unstoppable Owl

100) Word Problems

1) **Frank's bakery made 2,400 muffins. A third of these muffins were blueberry muffins. How many blueberry muffins did the bakery make?**

2) **Of the 2,400 muffins made, 200 were not sold by the end of the day. In simplest terms, what fraction of the muffins were not sold?**

3) **Each batch of chocolate chip muffins uses two and a fourth cups of chocolate chips. If the bakery makes 7 batches of chocolate chip muffins, how many cups of chocolate chips will be used?**

4) **The bakery had 13 and a half cartons of milk. Then, it used 4 and two-thirds cartons of milk. How many cartons of milk are left?** *State your answer as a mixed fraction in simplest form.*

5) **52 customers came to the bakery this morning. 9 of them ordered a slice of pie. Rounded to the nearest tenth, what percentage of the customers ordered pie this morning?**

6) **In the afternoon, 250 customers came to the bakery and 32% of them ordered pie. How many customers ordered pie in the afternoon?**

7) **If each sugar cookie contains 4.7 grams of fat, how many grams of fat will a package of 6 cookies contain?**

8) **If a group of 4 chocolate chip cookies contains 25.6 grams of fat, how many grams of fat does 1 chocolate chip cookie have?**

101) Answer Key

Page 1:

1) $\frac{2}{4}$ 2) $\frac{2}{3}$ 3) $\frac{1}{2}$ 4) $\frac{3}{5}$ 5) $\frac{1}{4}$

6) $\frac{2}{4}$ 7) $\frac{5}{6}$ 8) $\frac{1}{3}$ 9) $\frac{1}{5}$ 10) $\frac{2}{3}$

11) $\frac{2}{5}$ 12) $\frac{3}{4}$ 13) $\frac{5}{8}$ 14) $\frac{1}{6}$

15) $\frac{1}{2}$ 16) $\frac{6}{10}$ 17) $\frac{2}{4}$ 18) $\frac{4}{5}$

Page 2:

1) $\frac{1}{3}$ 2) $\frac{2}{5}$ 3) $\frac{1}{4}$ 4) $\frac{4}{8}$ 5) $\frac{1}{2}$

6) $\frac{3}{4}$ 7) $\frac{3}{5}$ 8) $\frac{2}{3}$ 9) $\frac{1}{2}$ 10) $\frac{2}{4}$

11) $\frac{5}{8}$ 12) $\frac{5}{6}$ 13) $\frac{3}{6}$ 14) $\frac{3}{4}$ 15) $\frac{2}{5}$

16) $\frac{1}{2}$ 17) $\frac{3}{4}$ 18) $\frac{4}{6}$ 19) $\frac{1}{8}$ 20) $\frac{4}{6}$

21) $\frac{1}{2}$ 22) $\frac{4}{5}$ 23) $\frac{7}{10}$ 24) $\frac{2}{3}$

Page 3:

1) $\frac{1}{2}$ 2) $\frac{1}{4}$ 3) $\frac{2}{4}$ 4) $\frac{4}{5}$ 5) $\frac{2}{3}$ 6) $\frac{2}{6}$

7) $\frac{3}{5}$ 8) $\frac{3}{6}$ 9) $\frac{7}{8}$ 10) $\frac{4}{6}$ 11) $\frac{1}{10}$

12) $\frac{5}{10}$ 13) $\frac{2}{3}$ 14) $\frac{2}{6}$ 15) $\frac{1}{5}$ 16) $\frac{3}{4}$

17) $\frac{6}{8}$ 18) $\frac{5}{6}$ 19) $\frac{2}{5}$ 20) $\frac{4}{10}$ 21) $\frac{1}{3}$

Page 4:

1) C 2) D 3) A 4) E 5) B
6) D 7) A 8) B 9) E 10) C
11) two-fourths 12) one-third
13) five-sixths 14) four-sevenths
15) one-fifth 16) three-tenths
17) five-eighths 18) three-fifths
19) two-ninths 20) seven-twelfths

21) $\frac{1}{6}$ 22) $\frac{3}{8}$ 23) $\frac{2}{5}$ 24) $\frac{5}{9}$ 25) $\frac{1}{2}$

26) $\frac{2}{7}$ 27) $\frac{6}{11}$ 28) $\frac{1}{10}$ 29) $\frac{7}{8}$ 30) $\frac{4}{5}$

Page 5:

1) $\frac{4}{6}$ 2) $\frac{3}{10}$ 3) $\frac{1}{4}$ 4) $\frac{2}{3}$ 5) $\frac{3}{4}$ 6) $\frac{3}{5}$

7) $\frac{6}{8}$ 8) $\frac{1}{6}$ 9) $\frac{1}{3}$ 10) $\frac{4}{5}$ 11) $\frac{4}{6}$ 12) $\frac{7}{10}$

13) six-sevenths 14) four-ninths
15) five-elevenths 16) two-thirds
17) one-fourth 18) two-sixths
19) nine-tenths 20) three-eighths
21) one-half 22) two-fifths

23) $\frac{3}{9}$ 24) $\frac{1}{3}$ 25) $\frac{5}{7}$ 26) $\frac{8}{10}$ 27) $\frac{3}{4}$

28) $\frac{1}{5}$ 29) $\frac{4}{8}$ 30) $\frac{3}{6}$ 31) $\frac{5}{12}$ 32) $\frac{1}{7}$

Page 6:

1) 1 part 2) 2 parts 3) 1 part
4) 7 parts 5) 3 parts 6) 2 parts
7) 3 parts 8) 1 part 9) 4 parts
10) 2 parts 11) 4 parts 12) 9 parts
13) 3 parts 14) 2 parts 15) 1 part
16) 5 parts 17) 7 parts 18) 5 parts
19) 3 parts 20) 1 part 21) 3 parts
22) 1 part 23) 4 parts 24) 5 parts

Page 7:

1) > 2) < 3) > 4) < 5) > 6) <
7) < 8) > 9) < 10) > 11) > 12) <
13) < 14) > 15) <

Page 8:

1) < 2) > 3) > 4) < 5) >
6) < 7) < 8) > 9) <

Page 9:

1) < 2) > 3) < 4) > 5) <
6) < 7) < 8) > 9) >

Page 10:

1) = 2) < 3) = 4) > 5) =
6) < 7) > 8) = 9) =

Page 11:

1) < 2) > 3) > 4) = 5) >
6) > 7) > 8) = 9) = 10) <
11) < 12) = 13) = 14) > 15) <

Page 12:

1) < 2) < 3) > 4) = 5) < 6) =
7) < 8) < 9) < 10) < 11) > 12) <
13) > 14) > 15) < 16) > 17) < 18) >

19) $\frac{3}{9}$ $\frac{1}{2}$ $\frac{2}{3}$ $\frac{5}{6}$ 20) $\frac{2}{10}$ $\frac{1}{3}$ $\frac{2}{4}$ $\frac{3}{5}$

Page 13:

1) $\frac{6}{15}$ 2) $\frac{4}{16}$ 3) $\frac{12}{14}$ 4) $\frac{5}{10}$ 5) $\frac{6}{9}$ 6) $\frac{6}{30}$

7) $\frac{8}{10}$ 8) $\frac{2}{3}$ 9) $\frac{3}{9}$ 10) $\frac{3}{8}$ 11) $\frac{30}{35}$ 12) $\frac{1}{2}$

Page 14:

1) $\frac{6}{8}$ 2) $\frac{8}{10}$ 3) $\frac{12}{16}$ 4) $\frac{20}{35}$ 5) $\frac{3}{4}$

6) $\frac{2}{6}$ 7) $\frac{15}{27}$ 8) $\frac{2}{10}$ 9) $\frac{16}{24}$ 10) $\frac{7}{10}$

11) $\frac{12}{36}$ 12) $\frac{2}{5}$ 13) $\frac{8}{12}$ 14) $\frac{3}{4}$ 15) $\frac{3}{10}$

16) $\frac{4}{14}$ 17) $\frac{8}{12}$ 18) $\frac{8}{18}$ 19) $\frac{4}{10}$ 20) $\frac{6}{8}$

Page 15:

1) 0.3 2) 0.8 3) 0.6 4) 0.2
5) 0.1 6) 0.4 7) 0.9 8) 0.5

Page 16:

1) 0.5 2) 0.2 3) 0.7 4) 0.9 5) 0.4
6) 0.1 7) 0.3 8) 0.8 9) 0.6 10) 0.4
11) 0.8 12) 0.1 13) 0.6 14) 0.9
15) 0.3 16) 0.7 17) 0.5

18) $\frac{2}{10}$ 19) $\frac{9}{10}$ 20) $\frac{6}{10}$ 21) $\frac{1}{10}$

22) $\frac{7}{10}$ 23) $\frac{5}{10}$ 24) $\frac{3}{10}$ 25) $\frac{8}{10}$

Page 17:

1) 0.13 2) 0.41 3) 0.75 4) 0.94
5) .07 6) 0.59 7) 0.88 8) 0.26

Page 18:

1) 0.29 2) 0.44 3) 0.02 4) 0.67 5) 0.95
6) 0.76 7) 0.51 8) 0.83 9) 0.18 10) 0.57
11) 0.11 12) 0.92 13) 0.35 14) 0.84
15) 0.49 16) 0.78 17) 0.03

18) $\frac{41}{100}$ 19) $\frac{28}{100}$ 20) $\frac{6}{100}$ 21) $\frac{94}{100}$

22) $\frac{72}{100}$ 23) $\frac{85}{100}$ 24) $\frac{53}{100}$ 25) $\frac{17}{100}$

Page 19:

1) 0.5 2) 0.25 3) 0.8 4) 0.6 5) 0.65 6) 0.85
7) 0.4 8) 0.84 9) 0.78 10) 0.4 11) 0.7 12) 0.15
13) Thousandths 14) Hundredths
15) Tenths 16) Ten Thousandths
17) Thousandths 18) Hundred Thousandths
19) Ten Thousandths 20) Ten Thousandths

Page 20:

1) 0.4, 0.37, 0.372, 0.3722 2) 0.9, 0.94, 0.936, 0.9357
3) 0.1, 0.13, 0.129, 0.1288 4) 0.6, 0.56, 0.556, 0.5556
5) 0.4, 0.37, 0.371, 0.3714 6) 0.8, 0.77, 0.773, 0.7729
7) 0.5, 0.48, 0.482, 0.4815 8) 0.9, 0.94, 0.942, 0.9418
9) 0.2, 0.22, 0.222, 0.2222 10) 0.6, 0.56, 0.562, 0.5619
11) 0.6, 0.62, 0.618, 0.6184 12) 0.9, 0.86, 0.863, 0.8627
13) 0.3, 0.31, 0.315, 0.3147 14) 0.8, 0.77, 0.769, 0.7686
15) 0.8, 0.84, 0.842, 0.8418

Page 21:

1) 0.38 2) 0.11 3) 0.22 4) 0.42 5) 0.07 6) 0.43 7) 0.53 8) 0.73
9) 0.65 10) 0.36 11) 0.44 12) 0.63 13) $0.\overline{8}$ 14) $0.\overline{47}$ 15) $0.2\overline{5}$
16) $0.5\overline{39}$ 17) $0.\overline{126}$ 18) $0.\overline{79}$ 19) $0.64\overline{3}$ 20) $0.\overline{81}$ 21) $0.\overline{2}$
22) $0.\overline{401}$ 23) $0.9\overline{47}$ 24) $0.\overline{582}$ 25) $0.\overline{6}$ 26) $0.9\overline{4}$ 27) $0.5\overline{4}$
28) $0.91\overline{6}$ 29) $0.\overline{185}$ 30) $0.13\overline{8}$ 31) $0.\overline{39}$ 32) $0.\overline{3}$ 33) $0.\overline{90}$
34) $0.34\overline{09}$ 35) $0.1\overline{36}$ 36) $0.\overline{381}$

Page 22:

1) 0.3 2) 0.4 3) 0.76 4) 0.081 5) 0.005 6) 0.2 7) 0.0409
8) 0.31 9) 0.2788 10) 0.7 11) 0.025 12) 0.0005 13) 0.8002
14) 0.261 15) 0.59 16) 0.00014 17) 0.94 18) 0.003 19) 0.387
20) 0.15 21) 0.020904 22) 0.7703 23) 0.0008 24) 0.01
25) 0.74 26) 0.51 27) 0.3 28) 0.1 29) 0.5 30) 0.94 31) 0.01
32) 0.13 33) 0.61 34) 0.58 35) 0.9 36) 0.05

Page 23:

1) 0.6 2) $0.\overline{3}$ 3) 0.4 4) 0.75 5) 0.5 6) 0.8 7) 0.7 8) 0.25 9) 0.1
10) 0.6 11) 0.5 12) 0.8 13) 0.2 14) 0.2 15) 0.5 16) 0.9 17) $0.\overline{6}$
18) 0.3 19) 0.4 20) 0.25 21) 0.5

Page 24:

1) 0.5 2) $0.\overline{3}$ 3) $0.\overline{6}$ 4) 0.25 5) 0.5 6) 0.75 7) 0.2
8) 0.4 9) 0.6 10) 0.8 11) 0.1 12) 0.2 13) 0.3 14) 0.4
15) 0.5 16) 0.6 17) 0.7 18) 0.8 19) 0.9

20) $\frac{1}{4}$ 21) $\frac{7}{10}$ 22) $\frac{4}{5}$ 23) $\frac{2}{10}$ 24) $\frac{9}{10}$ 25) $\frac{1}{3}$ 26) $\frac{2}{5}$

27) $\frac{5}{10}$ 28) $\frac{1}{10}$ 29) $\frac{3}{5}$ 30) $\frac{3}{4}$ 31) $\frac{4}{10}$ 32) $\frac{1}{2}$ 33) $\frac{2}{3}$

34) $\frac{3}{10}$ 35) $\frac{2}{4}$ 36) $\frac{8}{10}$ 37) $\frac{1}{5}$ 38) $\frac{6}{10}$ 39) $\frac{3}{4}$ 40) $\frac{1}{3}$

Page 25:

1) 0.2 2) 0.12 3) 0.12 4) 0.19 5) 0.45 6) 0.4
7) 0.28 8) 0.8 9) 0.07 10) 0.5 11) 0.9
12) 0.44 13) 0.68 14) 0.57 15) 0.06 16) 0.88
17) 0.75 18) 0.52 19) 0.87 20) 0.2 21) 0.36

Page 26:

1) 0.15 2) 0.2 3) 0.4 4) 0.7 5) 0.85 6) 0.16
7) 0.36 8) 0.56 9) 0.68 10) 0.96 11) 0.08
12) 0.14 13) 0.24 14) 0.5 15) 0.98
16) 0.06 17) 0.08 18) 0.33 19) 0.81

20) $\frac{21}{50}$ 21) $\frac{97}{100}$ 22) $\frac{12}{20}$ 23) $\frac{20}{25}$ 24) $\frac{29}{50}$

25) $\frac{7}{20}$ 26) $\frac{7}{25}$ 27) $\frac{63}{100}$ 28) $\frac{13}{20}$ 29) $\frac{22}{50}$

30) $\frac{7}{25}$ 31) $\frac{9}{50}$ 32) $\frac{3}{100}$ 33) $\frac{6}{20}$ 34) $\frac{5}{50}$

35) $\frac{8}{25}$ 36) $\frac{16}{20}$ 37) $\frac{17}{50}$ 38) $\frac{26}{100}$ 39) $\frac{23}{25}$ 40) $\frac{38}{50}$

102) Answer Key

Page 27:

1) 0.5 2) $0.\overline{3}$ 3) $0.\overline{6}$ 4) 0.25 5) 0.75
6) 0.6 7) 0.8 8) 0.1 9) 0.9 10) 0.05
11) 0.35 12) 0.08 13) 0.24 14) 0.48
15) 0.14 16) 0.46 17) 0.62 18) 0.29
19) 0.77 20) $\frac{7}{10}$ 21) $\frac{11}{20}$ 22) $\frac{2}{5}$ 23) $\frac{9}{25}$
24) $\frac{57}{100}$ 25) $\frac{2}{4}$ 26) $\frac{13}{50}$ 27) $\frac{4}{10}$ 28) $\frac{2}{3}$
29) $\frac{1}{25}$ 30) $\frac{1}{5}$ 31) $\frac{9}{100}$ 32) $\frac{1}{2}$ 33) $\frac{1}{3}$
34) $\frac{49}{50}$ 35) $\frac{15}{20}$ 36) $\frac{1}{4}$ 37) $\frac{2}{10}$ 38) $\frac{18}{20}$
39) $\frac{40}{50}$ 40) $\frac{4}{25}$

Page 28:

1) $1\frac{1}{2}$ 2) $5\frac{2}{5}$ 3) $7\frac{1}{4}$ 4) $3\frac{4}{6}$
5) $2\frac{2}{3}$ 6) $6\frac{1}{5}$ 7) $4\frac{1}{2}$ 8) $5\frac{5}{8}$

Page 29:

1) $5\frac{1}{3}$ 2) $3\frac{1}{5}$ 3) $6\frac{7}{8}$ 4) $8\frac{3}{4}$
5) $2\frac{2}{6}$ 6) $4\frac{3}{5}$ 7) $7\frac{2}{3}$ 8) $1\frac{4}{6}$
9) $5\frac{2}{4}$ 10) $2\frac{1}{2}$ 11) $8\frac{2}{5}$ 12) $4\frac{3}{4}$

Page 30:

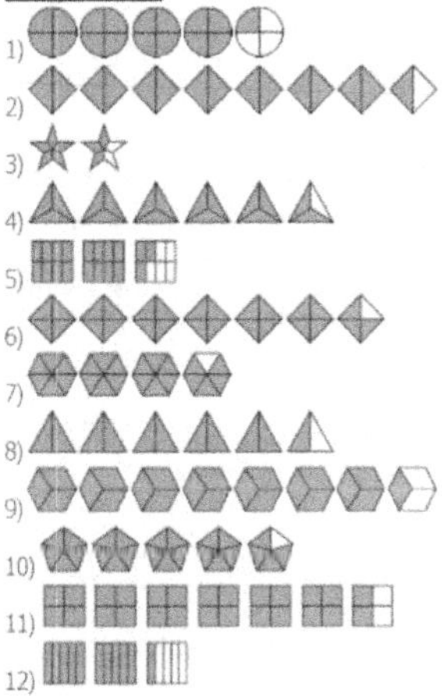

Page 31:

1) D 2) E 3) B 4) A 5) C
6) C 7) B 8) D 9) E 10) A
11) Six and five-eighths
12) Three and seven-ninths
13) Fifteen and fourty-one hundredths
14) Four and one half
15) Thirty-three and three-fifths
16) $1\frac{1}{11}$ 17) $13\frac{3}{8}$ 18) $7\frac{8}{9}$
19) $30\frac{33}{100}$ 20) $3\frac{7}{13}$

Page 32:

1) $4\frac{4}{5}$ 2) $7\frac{1}{2}$ 3) $2\frac{3}{4}$ 4) $5\frac{1}{6}$ 5) $6\frac{2}{3}$
6) Five and five-ninths
7) One hundred and one-third
8) Sixteen and three-elevenths
9) Two and four-sevenths
10) Eleven and one-sixth
11) Six and three-fourths

Page 32 Continued:

12) $4\frac{16}{25}$ 13) $24\frac{2}{5}$ 14) $18\frac{1}{3}$
15) $9\frac{10}{11}$ 16) $37\frac{7}{8}$ 17) $99\frac{1}{4}$

Page 33:

1) < 2) < 3) > 4) < 5) >
6) < 7) > 8) < 9) > 10) <

Page 34:

1) < 2) > 3) < 4) > 5) <
6) > 7) < 8) < 9) < 10) <

Page 35:

1) < 2) > 3) > 4) > 5) <
6) > 7) < 8) > 9) < 10) <
11) > 12) > 13) > 14) > 15) <
16) < 17) > 18) < 19) < 20) <
21) > 22) > 23) < 24) > 25) >
26) > 27) > 28) < 29) > 30) >

Page 36:

1) 5.75 2) 2.8 3) 9.7 4) 10.75
5) 1.67 6) 6.5 7) 4.4 8) 7.1
9) 3.05 10) 11.25 11) 5.9 12) 24.2
13) 9.03 14) 15.75 15) 32.4
16) 1.15 17) 7.33 18) 3.5
19) 6.06 20) 95.8 21) 47.67

Page 37:

1) > 2) < 3) > 4) < 5) = 6) >
7) < 8) = 9) > 10) > 11) < 12) <
13) < 14) < 15) > 16) < 17) <
18) > 19) < 20) = 21) >

Page 38:

1) $1\frac{1}{4}$, 1.5, 2.65, $2\frac{3}{4}$
2) 5.2, $5\frac{1}{4}$, 5.4, $5\frac{1}{2}$
3) 3.75, $3\frac{4}{5}$, 4.5, $4\frac{3}{5}$
4) $7\frac{4}{5}$, 7.85, $7\frac{9}{10}$, 8.2
5) $6\frac{6}{10}$, 6.65, $6\frac{2}{3}$, $6\frac{3}{4}$
6) $\frac{7}{10}$, 0.8, 1.5, $2\frac{3}{5}$
7) $9\frac{1}{5}$, 9.25, 9.4, 10.2
8) $3\frac{3}{10}$, 3.32, $3\frac{1}{3}$, 3.35
9) $5\frac{3}{5}$, $6\frac{1}{2}$, 6.6, 7.4
10) $2\frac{39}{100}$, $2\frac{2}{5}$, 2.44, 2.5
11) 10.41, $10\frac{9}{20}$, 11.4, $11\frac{1}{2}$
12) 5.55, $5\frac{2}{3}$, $6\frac{7}{10}$, $6\frac{3}{4}$
13) 6.8, $7\frac{2}{9}$, $7\frac{2}{7}$, 7.5
14) 2.05, $2\frac{1}{10}$, 2.2, $2\frac{1}{4}$
15) $4\frac{1}{5}$, 4.33, $4\frac{7}{20}$, 5.4

Page 39:

1) $\frac{8}{3}$ 2) $\frac{12}{5}$ 3) $\frac{3}{2}$ 4) $\frac{11}{6}$ 5) $\frac{19}{8}$
6) $\frac{11}{4}$ 7) $\frac{22}{6}$ 8) $\frac{7}{2}$ 9) $\frac{14}{4}$ 10) $\frac{9}{4}$
11) $\frac{31}{8}$ 12) $\frac{10}{3}$

Page 40:

1) $\frac{9}{4}$ 2) $\frac{16}{10}$ 3) $\frac{7}{3}$ 4) $\frac{15}{5}$ 5) $\frac{8}{4}$
6) $\frac{16}{6}$ 7) $\frac{12}{3}$ 8) $\frac{19}{5}$ 9) $\frac{24}{8}$ 10) $\frac{15}{4}$
11) $\frac{23}{6}$ 12) $\frac{12}{6}$

Page 41:

1) $\frac{23}{5}$ 2) $\frac{13}{4}$ 3) $\frac{58}{7}$ 4) $\frac{21}{2}$ 5) $\frac{37}{5}$
6) $\frac{28}{3}$ 7) $\frac{57}{5}$ 8) $\frac{51}{8}$ 9) $\frac{51}{4}$ 10) $\frac{37}{5}$
11) $\frac{35}{8}$ 12) $\frac{37}{6}$ 13) $\frac{43}{4}$ 14) $\frac{32}{11}$
15) $\frac{26}{3}$ 16) $\frac{13}{9}$ 17) $\frac{23}{4}$ 18) $\frac{9}{5}$
19) $\frac{87}{10}$ 20) $\frac{61}{3}$ 21) $\frac{39}{4}$ 22) $\frac{30}{7}$
23) $\frac{15}{2}$ 24) $\frac{71}{25}$

Page 42:

1) $2\frac{1}{2}$ 2) 3 3) $6\frac{3}{5}$ 4) $4\frac{1}{4}$
5) $5\frac{5}{8}$ 6) 9 7) $1\frac{2}{5}$ 8) $5\frac{7}{10}$
9) $4\frac{1}{3}$ 10) $11\frac{1}{2}$ 11) $6\frac{3}{4}$ 12) $7\frac{2}{9}$
13) $7\frac{4}{5}$ 14) $11\frac{1}{2}$ 15) 7 16) $4\frac{3}{8}$
17) $1\frac{6}{7}$ 18) $3\frac{7}{9}$ 19) $3\frac{4}{5}$ 20) $9\frac{1}{3}$
21) $10\frac{5}{7}$ 22) $11\frac{2}{5}$ 23) $3\frac{9}{10}$ 24) 8

Page 43:

1) $\frac{43}{7}$ 2) $\frac{19}{12}$ 3) $\frac{68}{6}$ 4) $\frac{49}{5}$ 5) $\frac{35}{8}$
6) $\frac{31}{2}$ 7) $\frac{94}{9}$ 8) $\frac{19}{6}$ 9) $\frac{31}{4}$ 10) $\frac{41}{7}$
11) $\frac{45}{4}$ 12) $\frac{53}{6}$ 13) $\frac{19}{3}$ 14) $\frac{77}{3}$
15) $\frac{47}{40}$ 16) $\frac{81}{8}$ 17) $\frac{84}{9}$ 18) $\frac{61}{25}$
19) $4\frac{2}{3}$ 20) $2\frac{7}{8}$ 21) 4 22) $5\frac{4}{5}$
23) $9\frac{1}{2}$ 24) 9 25) $5\frac{1}{7}$ 26) 4
27) $6\frac{2}{9}$ 28) $5\frac{3}{5}$ 29) $6\frac{7}{10}$
30) $11\frac{4}{5}$ 31) $3\frac{1}{12}$ 32) $20\frac{1}{2}$
33) $1\frac{5}{8}$ 34) $2\frac{13}{25}$ 35) $8\frac{7}{10}$ 36) 5

Page 44:

1) < 2) < 3) = 4) > 5) >
6) = 7) < 8) < 9) > 10) >
11) > 12) < 13) < 14) > 15) <
16) > 17) < 18) > 19) < 20) <
21) > 22) < 23) < 24) > 25) <
26) < 27) > 28) > 29) > 30) >

Page 45:

1) $\frac{3}{4}$ 2) $\frac{2}{7}$ 3) $\frac{2}{11}$ 4) $\frac{4}{5}$ 5) $\frac{1}{2}$ 6) $\frac{4}{5}$
7) $\frac{3}{5}$ 8) $\frac{1}{5}$ 9) $\frac{2}{5}$ 10) $\frac{3}{5}$ 11) $\frac{2}{3}$ 12) $\frac{2}{9}$
13) $\frac{4}{15}$ 14) $\frac{1}{3}$ 15) $\frac{9}{10}$ 16) $\frac{13}{22}$ 17) $\frac{1}{2}$
18) $\frac{2}{3}$ 19) $\frac{7}{8}$ 20) $\frac{14}{25}$ 21) $\frac{3}{4}$

Page 46:

1) $\frac{2}{5}$ 2) SF 3) SF 4) $\frac{5}{7}$ 5) $\frac{4}{11}$ 6) SF
7) SF 8) $\frac{7}{13}$ 9) SF 10) $\frac{5}{14}$ 11) SF 12) $\frac{8}{9}$
13) $\frac{1}{8}$ 14) $\frac{3}{4}$ 15) SF 16) $\frac{1}{7}$ 17) $\frac{5}{6}$ 18) SF
19) SF 20) $\frac{2}{5}$ 21) $\frac{17}{20}$ 22) SF 23) SF 24) $\frac{9}{16}$
25) $\frac{1}{3}$ 26) $\frac{4}{21}$ 27) $\frac{1}{5}$ 28) SF 29) $\frac{1}{2}$ 30) SF

Page 47:

1) $7\frac{1}{4}$ 2) $2\frac{2}{11}$ 3) $9\frac{12}{25}$ 4) $12\frac{2}{5}$
5) $5\frac{23}{30}$ 6) $8\frac{22}{25}$ 7) $10\frac{3}{10}$ 8) $1\frac{1}{3}$
9) $14\frac{9}{14}$ 10) $7\frac{18}{25}$ 11) $5\frac{2}{3}$ 12) $8\frac{8}{15}$
13) $17\frac{3}{5}$ 14) $4\frac{1}{3}$ 15) $6\frac{4}{9}$ 16) $3\frac{2}{9}$
17) $2\frac{1}{4}$ 18) $9\frac{3}{8}$ 19) $4\frac{4}{25}$ 20) $30\frac{3}{8}$
21) $1\frac{2}{13}$ 22) $7\frac{5}{8}$ 23) $5\frac{1}{8}$ 24) $8\frac{7}{11}$

Page 48:

1) $2\frac{1}{2}$ 2) $4\frac{1}{3}$ 3) $8\frac{3}{4}$ 4) $10\frac{1}{2}$ 5) $5\frac{1}{2}$
6) $3\frac{1}{3}$ 7) $3\frac{1}{2}$ 8) $6\frac{1}{5}$ 9) $2\frac{1}{7}$ 10) $1\frac{1}{2}$
11) $4\frac{1}{2}$ 12) $8\frac{1}{3}$ 13) $2\frac{1}{2}$ 14) $9\frac{1}{2}$
15) $2\frac{3}{5}$ 16) $4\frac{1}{5}$ 17) $2\frac{2}{3}$ 18) $5\frac{1}{10}$
19) $9\frac{1}{3}$ 20) $3\frac{3}{5}$ 21) $3\frac{4}{5}$ 22) $4\frac{2}{5}$
23) $4\frac{3}{4}$ 24) $1\frac{3}{4}$ 25) $4\frac{2}{5}$ 26) $14\frac{3}{5}$
27) $2\frac{5}{6}$ 28) $1\frac{3}{8}$ 29) $2\frac{4}{5}$ 30) $2\frac{3}{4}$

Page 49:

1) 17.77 2) 45.6 3) 132.88 4) 12.65
5) 4.99 6) 6.87 7) 5.609 8) 155.9
9) 27.2 10) 13.968 11) 58.9 12) 0.8
13) 7.47 14) 4.644 15) 181.5 16) 48.86

103) Answer Key

Page 50:

1) 6.54 2) 58.3 3) 43.03 4) 0.696
5) 75.03 6) 15.17 7) 13.1 8) 148.9
9) 4.17 10) 13.826 11) 162.4 12) 46.67
13) 2.1 14) 110.11 15) 7.628

Page 51:

1) 12.38 2) 83.24 3) 34.116 4) 10.187
5) 94.955 6) 0.8651 7) 11.186 8) 344.93
9) 38.31 10) 23.5538 11) 9.22261 12) 46.9219

Page 52:

1) 11.95 2) 9.12 3) 19.91 4) 87.29
5) 91.4 6) 170.5 7) 6.873 8) 437.2
9) 7.0508 10) 189.58 11) 62.4 12) 33.405

Page 53:

1) 11.06 2) 32.99 3) 110.78 4) 7.63
5) 10.309 6) 10.22 7) 8.576 8) 256.74
9) 6.018 10) 51.78 11) 1.225 12) 7.609
13) 927.5 14) 4.55 15) 69.3 16) 85.01
17) 382.71 18) 7.857

Page 54:

1) 2.9 2) 17.7 3) 5.049 4) 3.38
5) 0.39 6) 116.7 7) 28.45 8) 1.771
9) 1.363 10) 47.2 11) 5.48 12) 6.953

Page 55:

1) 16.7 2) 3.25 3) 56.58 4) 34.3
5) 0.562 6) 5.338 7) 150.8 8) 2.83
9) 5.61 10) 0.33 11) 4.852 12) 0.046
13) 56.75 14) 1.295 15) 228.13 16) 0.093
17) 8.587 18) 1.875

Page 56:

1) 4.7 2) 6.37 3) 6.218 4) 36.8
5) 12.15 6) 52.64 7) 3.68 8) 1.795
9) 0.742 10) 27.28 11) 16.7 12) 0.687
13) 3.7 14) 39.65 15) 0.055 16) 148.8
17) 10.63 18) 0.377

Page 57:

1) $\frac{7}{9}$ 2) $\frac{2}{3}$ 3) $\frac{8}{6}$ 4) $\frac{11}{2}$ 5) $\frac{8}{5}$ 6) $\frac{3}{8}$
7) $\frac{10}{15}$ 8) $\frac{12}{4}$ 9) $\frac{4}{10}$ 10) $\frac{14}{7}$ 11) $\frac{14}{2}$ 12) $\frac{9}{12}$
13) $\frac{16}{5}$ 14) $\frac{18}{7}$ 15) $\frac{5}{9}$ 16) $\frac{15}{20}$ 17) $\frac{35}{6}$ 18) $\frac{11}{14}$
19) $\frac{7}{3}$ 20) $\frac{21}{25}$ 21) $\frac{4}{30}$ 22) $\frac{8}{8}$ 23) $\frac{27}{4}$ 24) $\frac{13}{10}$
25) $\frac{9}{5}$ 26) $\frac{16}{100}$ 27) $\frac{3}{2}$ 28) $\frac{17}{6}$ 29) $\frac{70}{45}$ 30) $\frac{12}{18}$

Page 58:

1) $\frac{2}{6}$ 2) $\frac{1}{4}$ 3) $\frac{12}{9}$ 4) $\frac{2}{3}$ 5) $\frac{7}{12}$ 6) $\frac{14}{7}$
7) $\frac{7}{2}$ 8) $\frac{2}{5}$ 9) $\frac{4}{8}$ 10) $\frac{2}{10}$ 11) $\frac{16}{7}$ 12) $\frac{7}{3}$
13) $\frac{1}{20}$ 14) $\frac{5}{16}$ 15) $\frac{55}{5}$ 16) $\frac{7}{11}$ 17) $\frac{10}{9}$ 18) $\frac{2}{4}$
19) $\frac{1}{2}$ 20) $\frac{17}{35}$ 21) $\frac{23}{5}$ 22) $\frac{30}{70}$ 23) $\frac{3}{8}$ 24) $\frac{5}{12}$
25) $\frac{4}{6}$ 26) $\frac{32}{50}$ 27) $\frac{79}{7}$ 28) $\frac{53}{100}$ 29) $\frac{50}{28}$ 30) $\frac{1}{60}$

Page 59:

1) $4\frac{5}{6}$ 2) $8\frac{3}{4}$ 3) $8\frac{4}{7}$ 4) $7\frac{5}{10}$
5) $15\frac{4}{5}$ 6) $3\frac{2}{9}$ 7) $13\frac{5}{8}$ 8) $5\frac{3}{6}$
9) $5\frac{5}{11}$ 10) $9\frac{18}{20}$ 11) 12 12) $8\frac{1}{5}$
13) $17\frac{2}{7}$ 14) 6 15) $5\frac{7}{8}$ 16) $12\frac{6}{10}$
17) $6\frac{3}{4}$ 18) $7\frac{6}{7}$ 19) $1\frac{2}{3}$ 20) $4\frac{4}{16}$
21) $16\frac{7}{9}$ 22) $8\frac{13}{14}$ 23) $5\frac{1}{4}$ 24) 9
25) $6\frac{7}{20}$ 26) $31\frac{7}{10}$ 27) $9\frac{2}{4}$

Page 60:

1) $5\frac{5}{7}$ 2) 4 3) $8\frac{1}{8}$ 4) 9
5) $45\frac{1}{6}$ 6) $12\frac{6}{10}$ 7) 9 8) $3\frac{5}{7}$
9) $13\frac{6}{15}$ 10) $9\frac{2}{5}$ 11) $10\frac{7}{8}$ 12) $5\frac{5}{12}$
13) $8\frac{3}{5}$ 14) $11\frac{2}{9}$ 15) $6\frac{3}{5}$ 16) $11\frac{5}{7}$
17) $3\frac{12}{20}$ 18) $7\frac{4}{8}$ 19) $7\frac{2}{4}$ 20) $3\frac{7}{9}$
21) $8\frac{4}{7}$ 22) $28\frac{5}{25}$ 23) $8\frac{4}{6}$ 24) $13\frac{11}{15}$
25) $18\frac{8}{9}$ 26) $8\frac{7}{11}$ 27) 9

Page 61:

1) $67\frac{6}{7}$ 2) $18\frac{1}{6}$ 3) $44\frac{1}{3}$ 4) $14\frac{2}{5}$
5) $95\frac{3}{10}$ 6) $17\frac{1}{4}$ 7) $67\frac{7}{12}$ 8) $10\frac{5}{75}$
9) 8 10) $14\frac{6}{8}$ 11) $19\frac{9}{14}$ 12) $50\frac{1}{25}$
13) $55\frac{7}{9}$ 14) $15\frac{5}{11}$ 15) $3\frac{3}{7}$ 16) $31\frac{4}{9}$
17) $7\frac{4}{5}$ 18) 10 19) $45\frac{4}{17}$ 20) $12\frac{2}{15}$
21) $81\frac{1}{6}$ 22) $9\frac{26}{30}$ 23) $70\frac{3}{5}$ 24) $7\frac{9}{40}$
25) 30 26) $12\frac{7}{13}$ 27) 9
28) $47\frac{2}{3}$ 29) 6 30) $9\frac{6}{18}$

Page 62:

1) $2\frac{2}{5}$ 2) $3\frac{1}{7}$ 3) $1\frac{1}{2}$ 4) $4\frac{4}{5}$
5) $11\frac{2}{3}$ 6) $\frac{2}{3}$ 7) $4\frac{6}{10}$ 8) $5\frac{2}{3}$ 9) $\frac{3}{4}$
10) $2\frac{1}{3}$ 11) 3 12) $7\frac{1}{7}$ 13) $\frac{3}{8}$
14) $7\frac{5}{11}$ 15) $2\frac{2}{7}$ 16) $4\frac{7}{11}$ 17) $\frac{4}{15}$
18) $14\frac{1}{2}$ 19) $5\frac{7}{10}$ 20) 4 21) $3\frac{19}{30}$

Page 63:

1) $\frac{5}{8}$ 2) $\frac{5}{6}$ 3) $\frac{5}{6}$ 4) $\frac{1}{2}$ 5) $\frac{4}{5}$ 6) $\frac{2}{3}$
7) $\frac{3}{8}$ 8) $\frac{5}{7}$ 9) $\frac{7}{30}$ 10) $\frac{11}{12}$ 11) $\frac{17}{24}$ 12) $\frac{13}{16}$
13) $\frac{5}{9}$ 14) $\frac{16}{25}$ 15) $\frac{5}{8}$ 16) $\frac{7}{9}$ 17) $\frac{2}{3}$ 18) $\frac{3}{4}$
19) $\frac{5}{7}$ 20) $\frac{3}{4}$ 21) $\frac{9}{10}$ 22) $\frac{2}{5}$ 23) $\frac{11}{16}$ 24) $\frac{5}{6}$

Page 64:

1) $\frac{11}{12}$ 2) $\frac{11}{15}$ 3) $\frac{5}{6}$ 4) $\frac{5}{8}$ 5) $\frac{19}{30}$ 6) $\frac{16}{15}$
7) $\frac{13}{14}$ 8) $\frac{7}{12}$ 9) $\frac{17}{20}$ 10) $\frac{23}{24}$ 11) $\frac{19}{28}$ 12) $\frac{5}{6}$
13) $\frac{9}{10}$ 14) $\frac{13}{18}$ 15) $\frac{17}{20}$ 16) $\frac{7}{12}$ 17) $\frac{7}{9}$ 18) $\frac{11}{15}$
19) $\frac{13}{18}$ 20) $\frac{11}{15}$ 21) $\frac{19}{24}$

Page 65:

1) $\frac{25}{8}$ 2) $\frac{25}{6}$ 3) $\frac{29}{9}$ 4) $\frac{27}{10}$ 5) $\frac{43}{12}$ 6) $\frac{67}{24}$
7) $\frac{21}{10}$ 8) $\frac{47}{14}$ 9) $\frac{19}{6}$ 10) $\frac{49}{20}$ 11) $\frac{43}{24}$ 12) $\frac{7}{2}$
13) $\frac{31}{18}$ 14) $\frac{13}{3}$ 15) $\frac{23}{4}$ 16) $\frac{41}{12}$ 17) $\frac{31}{5}$ 18) $\frac{101}{35}$
19) $\frac{47}{10}$ 20) $\frac{125}{44}$ 21) $\frac{35}{6}$ 22) $\frac{89}{30}$ 23) $\frac{7}{3}$ 24) $\frac{121}{36}$
25) $\frac{45}{8}$ 26) $\frac{25}{6}$ 27) $\frac{10}{3}$

Page 66:

1) $20\frac{1}{8}$ 2) $16\frac{4}{15}$ 3) $32\frac{7}{12}$ 4) 4
5) $7\frac{7}{30}$ 6) $6\frac{1}{18}$ 7) $40\frac{2}{15}$ 8) $5\frac{8}{9}$
9) $6\frac{9}{10}$ 10) $8\frac{23}{28}$ 11) $15\frac{11}{20}$ 12) 100
13) $8\frac{29}{30}$ 14) $5\frac{7}{24}$ 15) $66\frac{5}{36}$ 16) $6\frac{11}{15}$
17) $8\frac{5}{24}$ 18) 79 19) $24\frac{23}{30}$ 20) $2\frac{35}{36}$
21) $35\frac{5}{21}$ 22) $11\frac{7}{18}$ 23) $15\frac{3}{4}$ 24) $9\frac{1}{15}$
25) $5\frac{5}{28}$ 26) $54\frac{5}{8}$ 27) $8\frac{17}{24}$

104) Answer Key

Page 67:

1) $\frac{6}{7}$ 2) $9\frac{1}{3}$ 3) $1\frac{2}{3}$ 4) $14\frac{7}{24}$ 5) $6\frac{3}{5}$

6) $9\frac{11}{15}$ 7) $60\frac{7}{12}$ 8) $9\frac{13}{35}$ 9) 7 10) $7\frac{1}{24}$

11) $1\frac{2}{9}$ 12) $8\frac{1}{2}$ 13) $2\frac{2}{3}$ 14) $12\frac{1}{10}$ 15) $15\frac{2}{3}$

16) $1\frac{4}{7}$ 17) $19\frac{29}{36}$ 18) $5\frac{9}{10}$ 19) $1\frac{1}{24}$ 20) $28\frac{1}{3}$

21) $8\frac{1}{10}$ 22) $1\frac{2}{7}$ 23) $75\frac{7}{10}$ 24) 7 25) $48\frac{1}{2}$

26) $7\frac{9}{40}$ 27) $\frac{11}{12}$ 28) $5\frac{7}{8}$ 29) $9\frac{19}{40}$ 30) $8\frac{2}{3}$

31) $2\frac{13}{20}$ 32) $9\frac{2}{9}$ 33) $1\frac{1}{24}$

Page 68:

1) $\frac{1}{8}$ 2) $\frac{2}{7}$ 3) $\frac{7}{15}$ 4) $\frac{1}{28}$ 5) $\frac{5}{18}$ 6) $\frac{3}{10}$

7) $\frac{5}{16}$ 8) $\frac{3}{8}$ 9) $\frac{4}{35}$ 10) $\frac{2}{25}$ 11) $\frac{3}{20}$ 12) $\frac{2}{9}$

13) $\frac{8}{21}$ 14) $\frac{3}{10}$ 15) $\frac{17}{40}$ 16) $\frac{1}{2}$ 17) $\frac{1}{4}$ 18) $\frac{23}{35}$

19) $\frac{17}{45}$ 20) $\frac{1}{24}$ 21) $\frac{13}{40}$ 22) $\frac{9}{28}$ 23) $\frac{1}{10}$ 24) $\frac{7}{30}$

25) $\frac{3}{8}$ 26) $\frac{7}{18}$ 27) $\frac{1}{6}$

Page 69:

1) $2\frac{7}{20}$ 2) $4\frac{13}{14}$ 3) $\frac{7}{12}$ 4) $5\frac{1}{15}$

5) $4\frac{8}{9}$ 6) $2\frac{23}{30}$ 7) $5\frac{1}{5}$ 8) $\frac{17}{20}$

9) 7 10) $1\frac{2}{3}$ 11) $1\frac{5}{6}$ 12) $6\frac{29}{40}$

13) 3 14) $14\frac{19}{24}$ 15) $2\frac{1}{9}$ 16) $7\frac{11}{20}$

17) $4\frac{3}{16}$ 18) $3\frac{3}{20}$ 19) $2\frac{17}{21}$ 20) $2\frac{4}{5}$

21) $\frac{11}{15}$ 22) $37\frac{7}{15}$ 23) $6\frac{4}{15}$ 24) $4\frac{29}{36}$

25) $2\frac{1}{25}$ 26) $2\frac{7}{9}$ 27) $9\frac{15}{28}$

Page 70:

1) $1\frac{2}{3}$ 2) $\frac{6}{35}$ 3) $\frac{3}{4}$ 4) $6\frac{17}{24}$

5) $4\frac{17}{30}$ 6) $1\frac{1}{6}$ 7) $\frac{3}{13}$ 8) $\frac{2}{5}$

9) $3\frac{1}{3}$ 10) $24\frac{33}{40}$ 11) $4\frac{1}{36}$ 12) $\frac{5}{24}$

13) $7\frac{1}{12}$ 14) $\frac{5}{6}$ 15) $3\frac{23}{30}$ 16) $5\frac{11}{12}$

17) $\frac{3}{10}$ 18) $1\frac{1}{2}$ 19) $8\frac{2}{7}$ 20) $1\frac{1}{2}$

21) $2\frac{35}{44}$ 22) $\frac{9}{20}$ 23) $\frac{1}{6}$ 24) $7\frac{4}{7}$

25) $3\frac{5}{24}$ 26) $3\frac{7}{12}$ 27) $1\frac{13}{16}$ 28) $4\frac{7}{9}$

29) $9\frac{7}{8}$ 30) 2 31) $\frac{5}{18}$ 32) $3\frac{7}{9}$

33) $\frac{14}{15}$ 34) $2\frac{11}{15}$ 35) $6\frac{5}{8}$ 36) $\frac{7}{10}$

Page 71:

1) 48% 2) 31% 3) 20% 4) 9%
5) 74% 6) 263% 7) 1% 8) 508%
9) 99% 10) 0.2% 11) 137% 12) 56%
13) 310.5% 14) 600% 15) 40% 16) 0.7%
17) 105.2% 18) 3% 19) 260% 20) 91.4%
21) 22% 22) 80% 23) 16% 24) 703%
25) 0.1% 26) 40.5% 27) 55% 28) 100.3%
29) 60% 30) 4% 31) 6.1% 32) 825%
33) 400% 34) 600.1% 35) 0.45% 36) 10%
37) 0.08% 38) 1,250% 39) 370% 40) 20.75%

Page 72:

1) 0.83 2) 0.09 3) 0.3 4) 4.75
5) 0.007 6) 2 7) 0.058 8) 1.06
9) 0.0005 10) 0.11 11) 0.076 12) 7.2
13) 0.55 14) 0.004 15) 1.5 16) 0.1
17) 0.03 18) 0.605 19) 0.0814 20) 0.9
21) 0.0039 22) 0.08 23) 0.71 24) 1.11
25) 0.05 26) 0.0009 27) 0.042 28) 0.504
29) 0.0295 30) 2.6 31) 0.008 32) 0.067
33) 0.45 34) 0.371 35) 10 36) 0.6
37) 0.02003 38) 0.085 39) 0.00172 40) 8
41) 25 42) 41.95 43) 0.00063 44) 5.062

Page 73:

1) 30% 2) 4% 3) 225% 4) 73%
5) 545% 6) 50% 7) 14% 8) 170%
9) 80% 10) 92% 11) 625% 12) 95%
13) 6% 14) 2050% 15) 80% 16) 15%
17) 44% 18) 39% 19) 360% 20) 7%
21) 44% 22) 56.25% 23) 405% 24) 13.75%
25) 60% 26) 112.5% 27) 47.5% 28) 218%
29) 37.5% 30) 5.75% 31) 2.5% 32) 0.5%
33) 0.7% 34) 775% 35) 6.25% 36) 1.75%
37) 350% 38) 6.3% 39) 1020% 40) 18.05%

Page 74:

1) 60% 2) 25% 3) 50% 4) 90%
5) 80% 6) 75% 7) 50% 8) 66.67%
9) 70% 10) 50% 11) 40% 12) 25%
13) 80% 14) 20% 15) 33.33% 16) 50%
17) 66.67% 18) 10%

Page 75:

1) 2 parts 2) 1 part 3) 1 part
4) 3 parts 5) 3 parts 6) 4 parts
7) 2 parts 8) 7 parts 9) 2 parts
10) 7 parts 11) 3 parts 12) 8 parts
13) 2 parts 14) 6 parts 15) 6 parts
16) 4 parts 17) 4 parts 18) 1 part

Page 76:

1) 12 2) 6 3) 10 4) 15
5) 3 6) 15 7) 5 8) 4

Page 77:

1) 3,258 2) 267 3) 168 4) 40
5) 36 6) 1,015 7) 160 8) 90
9) 600 10) 1,200 11) 99 12) 5
13) 1,584 14) 5 15) 259 16) 7,500
17) 225 18) 198 19) 348 20) 80

Page 78:

1) 31.2 2) 43.75 3) 22 4) 18.3

Page 79:

1) 10.6 2) 8.16 3) 55.5 4) 12.8
5) 94.5 6) 43.2 7) 3.76 8) 10.38
9) 840 10) 121.5 11) 47 12) 3
13) 60.8 14) 18.76 15) 2.4 16) 22.8
17) 6.5 18) 16.5 19) 24.6 20) 74.76

Page 80:

1) 14.31 2) 3.249 3) 33.21 4) 0.12
5) 3.7512 6) 7.308 7) 0.4774 8) 26.52

Page 81:

1) 1.435 2) 58.93 3) 8.3214 4) 25.28
5) 9.4905 6) 0.8848 7) 14.067 8) 13.2046
9) 8.624 10) 7.05 11) 20.565 12) 3.30495
13) 2.01 14) 0.0688 15) 6.516 16) 5.5507

Page 82:

1) 280 2) 5.6 3) 25.6 4) 1.75
5) 216 6) 9 7) 150 8) 4.95
9) 0.06 10) 259

Page 83:

1) 1.5 2) 0.2 3) 4.13 4) 0.24
5) 2.4 6) 5.9 7) 0.35 8) 1.26

Page 84:

1) 1.66 2) 0.035 3) 0.575 4) 12.65
5) 0.2175 6) 0.12 7) 1.895 8) 0.535

Page 85:

1) 0.55 2) 0.075 3) 1.79 4) 1.839
5) 0.99 6) 16.45 7) 8.7 8) 0.0725
9) 4.5 10) 0.975 11) 0.006 12) 0.72
13) 0.945 14) 0.61 15) 4.675 16) 1.16

Page 86:

1) 0.12 2) 0.51 3) 4.68 4) 1.49
5) 27.87 6) 0.15 7) 0.89 8) 0.48

Page 87:

1) $\frac{14}{15}$ 2) $4\frac{2}{3}$ 3) 34 4) $1\frac{9}{16}$ 5) $\frac{16}{25}$

6) $\frac{5}{9}$ 7) $3\frac{2}{3}$ 8) $6\frac{3}{7}$ 9) $2\frac{2}{3}$ 10) $\frac{25}{48}$

11) $5\frac{1}{4}$ 12) $2\frac{1}{2}$ 13) $9\frac{1}{6}$ 14) $3\frac{1}{2}$

Page 88:

1) $4\frac{4}{15}$ 2) $1\frac{11}{25}$ 3) $\frac{1}{12}$ 4) 1 5) $6\frac{4}{25}$

6) $2\frac{1}{7}$ 7) 30 8) $\frac{1}{3}$ 9) $\frac{8}{9}$ 10) $1\frac{3}{4}$ 11) $7\frac{1}{3}$

12) 25 13) $15\frac{3}{4}$ 14) $4\frac{1}{4}$ 15) $\frac{11}{12}$ 16) $4\frac{3}{8}$

17) $\frac{2}{3}$ 18) $1\frac{1}{3}$ 19) $6\frac{3}{10}$ 20) $13\frac{1}{3}$ 21) $6\frac{1}{4}$

22) $\frac{3}{5}$ 23) $\frac{5}{7}$ 24) $14\frac{2}{5}$

Page 89:

1) $\frac{5}{27}$ 2) $9\frac{1}{3}$ 3) 36 4) $8\frac{1}{6}$ 5) $4\frac{3}{5}$ 6) 2

7) $4\frac{7}{12}$ 8) $\frac{9}{16}$ 9) $1\frac{4}{5}$ 10) $28\frac{1}{6}$ 11) 14

12) $1\frac{9}{16}$ 13) $\frac{1}{4}$ 14) $1\frac{3}{5}$ 15) $1\frac{2}{7}$ 16) $8\frac{1}{10}$

17) $7\frac{1}{2}$ 18) $\frac{1}{2}$ 19) $1\frac{1}{14}$ 20) $14\frac{7}{10}$ 21) $1\frac{1}{24}$

22) $22\frac{1}{2}$ 23) $\frac{1}{72}$ 24) $\frac{21}{32}$

105) Answer Key

Page 90:

1) $\frac{3}{5}$ 2) $\frac{4}{7}$ 3) $2\frac{6}{7}$ 4) $\frac{3}{35}$ 5) $4\frac{4}{9}$

6) $1\frac{7}{25}$ 7) $1\frac{13}{27}$ 8) $1\frac{13}{35}$ 9) $\frac{5}{32}$ 10) $6\frac{2}{3}$

11) $2\frac{1}{16}$ 12) $\frac{5}{8}$ 13) $2\frac{7}{10}$ 14) $3\frac{1}{3}$

Page 91:

1) $\frac{1}{4}$ 2) $1\frac{11}{24}$ 3) $\frac{7}{16}$ 4) $\frac{7}{10}$ 5) 6 6) $2\frac{5}{8}$

7) $4\frac{1}{5}$ 8) $11\frac{1}{4}$ 9) $\frac{4}{27}$ 10) $1\frac{1}{3}$ 11) 6 12) $4\frac{2}{3}$

13) $\frac{35}{44}$ 14) $6\frac{3}{7}$ 15) 2 16) $1\frac{13}{32}$ 17) 16 18) $\frac{1}{12}$

19) $\frac{14}{27}$ 20) $\frac{3}{10}$ 21) 11 22) $\frac{21}{32}$ 23) $1\frac{9}{11}$ 24) $5\frac{1}{3}$

Page 92:

1) 7 2) $17\frac{1}{2}$ 3) $\frac{27}{80}$ 4) $1\frac{7}{15}$ 5) $7\frac{1}{2}$ 6) $2\frac{3}{4}$

7) $\frac{4}{9}$ 8) $3\frac{3}{5}$ 9) $3\frac{3}{10}$ 10) $\frac{15}{16}$ 11) $12\frac{1}{2}$ 12) $\frac{9}{50}$

13) $\frac{7}{15}$ 14) $\frac{16}{45}$ 15) $\frac{1}{9}$ 16) 9 17) 1 18) $\frac{1}{21}$

19) $2\frac{3}{4}$ 20) $1\frac{19}{21}$ 21) $6\frac{2}{5}$ 22) $\frac{1}{7}$ 23) $3\frac{1}{3}$ 24) 48

Page 93:

1) 5.39 2) 160.1 3) 10.701 4) 1.332
5) 6.37 6) 1.769 7) 17.94 8) 0.665
9) 23.606 10) 1.22 11) 16.328 12) 8.6275
13) 0.83 14) 0.94 15) 0.26 16) 1.425

Page 94:

1) $8\frac{13}{14}$ 2) $6\frac{2}{5}$ 3) $5\frac{1}{2}$ 4) $6\frac{5}{12}$ 5) $3\frac{29}{40}$ 6) $1\frac{5}{18}$

7) $7\frac{1}{8}$ 8) $3\frac{17}{30}$ 9) $6\frac{13}{20}$ 10) $2\frac{9}{10}$ 11) $2\frac{1}{3}$ 12) $3\frac{7}{20}$

13) $\frac{5}{6}$ 14) $3\frac{11}{12}$ 15) $7\frac{23}{24}$ 16) $\frac{19}{30}$ 17) $8\frac{1}{9}$ 18) $\frac{5}{12}$

19) $1\frac{13}{15}$ 20) $\frac{1}{2}$ 21) $2\frac{11}{12}$ 22) $13\frac{1}{3}$ 23) $38\frac{1}{2}$ 24) $\frac{1}{4}$

25) $\frac{2}{9}$ 26) $6\frac{1}{2}$ 27) $5\frac{5}{6}$ 28) $\frac{3}{20}$ 29) $3\frac{3}{4}$ 30) $\frac{4}{5}$

Page 95:

1) 78% 2) 259% 3) 12% 4) 7%
5) 64% 6) 0.2% 7) 90% 8) 140%
9) 30% 10) 470.5% 11) 1.6% 12) 308%
13) 0.27 14) 0.03 15) 1.85 16) 0.007
17) 0.056 18) 4 19) 0.704 20) 15
21) 0.0061 22) 0.028 23) 0.0006 24) 0.05
25) 80% 26) 25% 27) 50% 28) 100%
29) 40% 30) 50% 31) 140 32) 4
33) 90 34) 75 35) 310 36) 36
37) 225 38) 5 39) 520 40) 98

Page 96:

1) 10,400 steps (13,000 × 0.8)
2) 14,760 steps (12,000 + (12,000 × 0.23))
3) 10,200 steps (12,000 – (12,000 × 0.15)
4) 1.82 km (7.2 – 5.38)
5) 27 push-ups (18 + (18 × 0.5))
6) 150% (Increase = 15 – 6 = 9.
Percent increase = 9 ÷ 6 = 1.5 = 150%
7) 15 minutes (45 ÷ 3)
8) 39 minutes (52 × 0.75)

Page 97:

1) 1,956 customers (1,630 + (1,630 × 0.2))
2) 145 Caesar Salads (580 × 0.25)
3) 945 large plates (270 × 3.5)
4) 913 cups (1,100 – (1,100 × 0.17))
5) 572 cups (1,430 × 0.4)
6) 70% (140 ÷ 200 = 0.7 = 70%)
7) 8 1/4 bags (2 3/4 × 3)
8) 2/5 napkin boxes (16/40 = 2/5)

Page 98:

1) 1 1/12 bins (1/3 + 3/4)
2) 7 1/3 boxes (2/3 × 11)
3) 152 nails (400 × 0.38)
4) 8 1/4 boxes (4 1/2 + 3 3/4)
5) 116.22 cm (19.37 × 6)
6) 76.7 cm (94.2 – 17.5)
7) 42 sections (5.25 × 8)
8) 2/7 (8/28 = 2/7)

Page 99:

1) 0.72 sec (50 – 49.28)
2) 6.13 sec (52.3 – 46.17)
3) 113.2 sec (1 min 53.2 sec) (28.3 × 4)
4) 30.41 sec (121.64 ÷ 4)
5) 7.5% (80 ÷ 6 = 0.075 = 7.5%)
6) 1/5 (16/80 = 1/5)
7) 4.45 sec (39.05 – 34.6)
8) 42.86 sec (43.11 – 0.25)

Page 100:

1) 800 blueberry muffins (2,400 × 1/3)
2) 1/12 (200/2,400 = 1/12)
3) 15 3/4 cups (7 × 2 1/4)
4) 8 5/6 cartons (13 1/2 - 4 2/3)
5) 17.3% (9 ÷ 52)
6) 80 customers (250 × 0.32)
7) 28.2 grams (4.7 × 6)
8) 6.4 grams (25.6 ÷ 4)

ISBN: 978-1-63578-555-5
Current contact information can be found at:
www.UnstoppableOwl.com www.LibroStudioLLC.com

Made in United States
North Haven, CT
10 November 2026

22106133R00059